Introductory Engineering Series

Under the Editorship of G. A. Webster, BSc(Eng), PhD, DIC

Thermodynamic cycles and processes

Russell Hoyle, *BA Camb., PhD Lond., FIMechE, Emeritus Professor of Engineering Science, University of Durham*
and
P. H. Clarke, *BSc, PhD Hull, Lecturer in Engineering Science, University of Durham*

Longman

LONGMAN GROUP LIMITED
London

*Associated companies, branches and representatives
throughout the world*

© Longman Group Limited 1973

First published 1973
ISBN 0582 44202. 8 cased
ISBN 0582 44203. 6 paper

PRINTED BY Unwin Brothers Limited
THE GRESHAM PRESS OLD WOKING SURREY ENGLAND

Produced by 'Uneoprint'
A member of the Staples Printing Group

Preface

Thermodynamics is part of the science behind much of life, ranging from the laws governing genetic behaviour to simple problems such as 'Will a jelly solidify?' Engineering thermodynamics considers a limited, but still large, range of problems from the boiling of water to space travel. These actions include as an essential part: heating as when boiling water, working as in raising the space vehicle from the Earth and the changes that occur when the water boils and the fuel loss of the space vehicle.

Most of the processes mentioned above are to some extent carried out by men of many nations speaking different languages. It is not surprising therefore that while some spoke of 'British Thermal Units', these to others were 'Calories'; or what was horsepower to some was kilowatts to others. Even in one language the variety was enormous. A British 1934 diary quotes *hand, span, inch, foot, yard, pace, fathom, rod, chain, furlong* and *mile* as permissible units of length. It would have been with nothing more than sentimental regrets that we replaced this plethora of units by multiples of the *metre*. However, such simplification did not satisfy us and although we got rid of the *pennyweight, dram, bushel, pottle, chaldron,* and *wey* we have gained or retained such oddities as *newtons, bars, pascals, hertz, watts* and others. In this book it would have been quixotic of us to have adhered uncompromisingly to the use of the *metre, kilogramme, second* and *Kelvin* and so we have used the variations allowed by the Système International.

One of the most stubborn obstacles to clear thinking in thermodynamics is the double meaning of heat, handed down from the caloric theory. One says that one heats an object by transferring heat to it. In fact people frequently assert that they transfer heat by heating. Joule pointed out that the property transferred by work was, after transfer, indistinguishable from the property transferred by heat. Those who have to choose their words carefully to maintain clarity now use three words. Two are actions of transferring—to *heat* and to *work*. The other is *energy*, the thing transferred (always realising there are many forms of energy—kinetic and potential energy being two of the simpler forms).

Much of our effort in writing this book has been given to simplification and clarity. In all our efforts we are indebted to others whose opposition to a simpler approach has made us think carefully before committing ourselves.

Our thanks are also due to Miss Carole Anderson for her accuracy in preparing the manuscript and to Mr Jim Moseley for his careful preparation of the outline figures and graphs.

References

1. Haywood, R. W. *Analysis of Engineering Cycles*, Pergamon Press, London, 1967.

2. Rogers, G. F. C. and Mayhew, Y. R. *Engineering Thermodynamics Work and Heat Transfer*, Longman, London, 1967.

3. Mayhew, Y. R. and Rogers, G. F. C. *Thermodynamic and Transport Properties of Fluids*, Blackwell, Oxford, 1971.

4. Hickson, D. C. and Taylor, F. R. *Entropy Diagram for Steam*, Blackwell, Oxford.

Contents

1

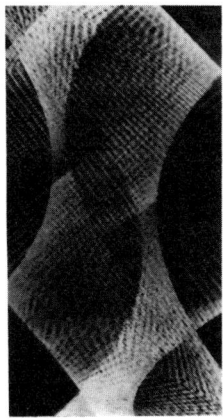

What thermodynamics is about

In order to understand what thermodynamics is and why it is worth studying we must look at Man against the background of the world in which he lives. Man is a relatively weak creature. The power that he can develop when compared with that of his surroundings, for example wind and tide, demonstrates Man's insecure position. Although in technological societies Man manages to create the illusion of being secure against the whims of nature he still spends his life using his brain to utilise these potentially dangerous forces for his own benefit. Man is capable of little sustained effort—about equal to that of a small electric motor of 0.05 kW—and he can lose energy to his surroundings by radiation, convection and sweating at the rate of 0.25 kW—about four people equal a one-bar electric fire. These powers must be compared with power stations which man has designed and which generate in the order of one million kW of power. This book describes the fundamental components and basic methods of analysis of various power producing plants.

1.1   Power plants

In early days man used animals for power production. A horse for example is capable of a greater sustained effort than man (0.75 kW). Later animals were supplemented by windmills and waterwheels capable of developing in the range of 1-6 kW of power. However, the advent of the industrial revolution caused the engineer to search for larger and more reliable sources of power not dependent on climatic conditions. His search led him first to fossil and then to nuclear fuels. To convert the energy stored in these fuels into useful energy capable of doing work he invented a device called a heat-engine. There are many types of heat-engine depending on the fuel used and the practical purpose to which the engine is to be put. The mode of operation of these heat-engines is discussed in later chapters of the book.

Today we have an increasing additional problem in the increasing demands for power due to a general increase in the standard of living and an increase in world population. Indigenous fuels, such as fossil and nuclear fuels although abundant originally, will become scarce if power consumption continues to increase at the rate predicted. Future developments may harness solar energy directly and result in the development of yet further heat-engines.

1.2   Heat and work

Thermodynamics is the fundamental science governing the processes which occur in a heat engine. A useful definition of thermodynamics is *a study of the relationship between heat, work and the stored energy contained by the matter under consideration*. Thermodynamics is concerned with energy balances and with explaining the direction in which a process will go. In its purest form thermodynamics is an exact science and, using calculation and logic, it may assist the engineer when designing heat engines. It will help him decide the size of components for specific power requirements and will answer such questions as, 'If I change a component in the heat engine how will this change its performance?' Or, 'Why did the last change I made affect performance in the particular way that it was seen to do?'

1.3   Simple processes

Engineers design and build heat engines to supplement Man's strength. There are many forms of these ranging from the giant power plants of the electricity authorities and of spacecraft to medium-sized power plants that drive ships, to small power plants that may form part of artificial hearts, (see Figs. 1.1, 1.2, 1.3 and 1.4). They all, however, have one thing in common—each takes in energy from a readily available source and gives out energy by work. The nuclear power plant takes energy liberated by nuclear reactions, the space-craft from the chemical reaction of solid or liquid fuel, the engine of a ship takes in energy perhaps from oil, and the artificial heart takes in power from a small nuclear source. The nuclear power plant gives out energy in the form of electric power, the space-craft does work primarily against gravity, the engine of the ship by work on the sea via the propellor, and the artificial heart does work by pumping the blood around the body.

Power units that take energy in by heat and give out energy by work are heat engines in the thermodynamic sense. All heat engines are represented in their simplest form by Fig. 1.5, which is a box E taking in energy Q_1 by heat and giving out energy W by work. As we will see later in the book they cannot operate without also giving out some energy Q_2 by heat. Energy in a form suitable for transfer by heat into a heat engine can be found in coal, oil, gas and from radioactive sources. The release of chemical energy generally causes a rise of temperature in the products of combustion. The high temperature of these combustion products if brought into contact with a conducting colder surface will cause an energy transfer by heat to occur. This energy Q_1 is shown entering the heat engine E in Fig. 1.5.

FIG 1. 1 (a)   Dungeness 'B' nuclear power station in Kent.

FIG 1. 1 (b)   A partly dismantled turbine at Dounreay nuclear power
station. *(Photographs by courtesy of the United Kingdom
Atomic Energy Authority.)*

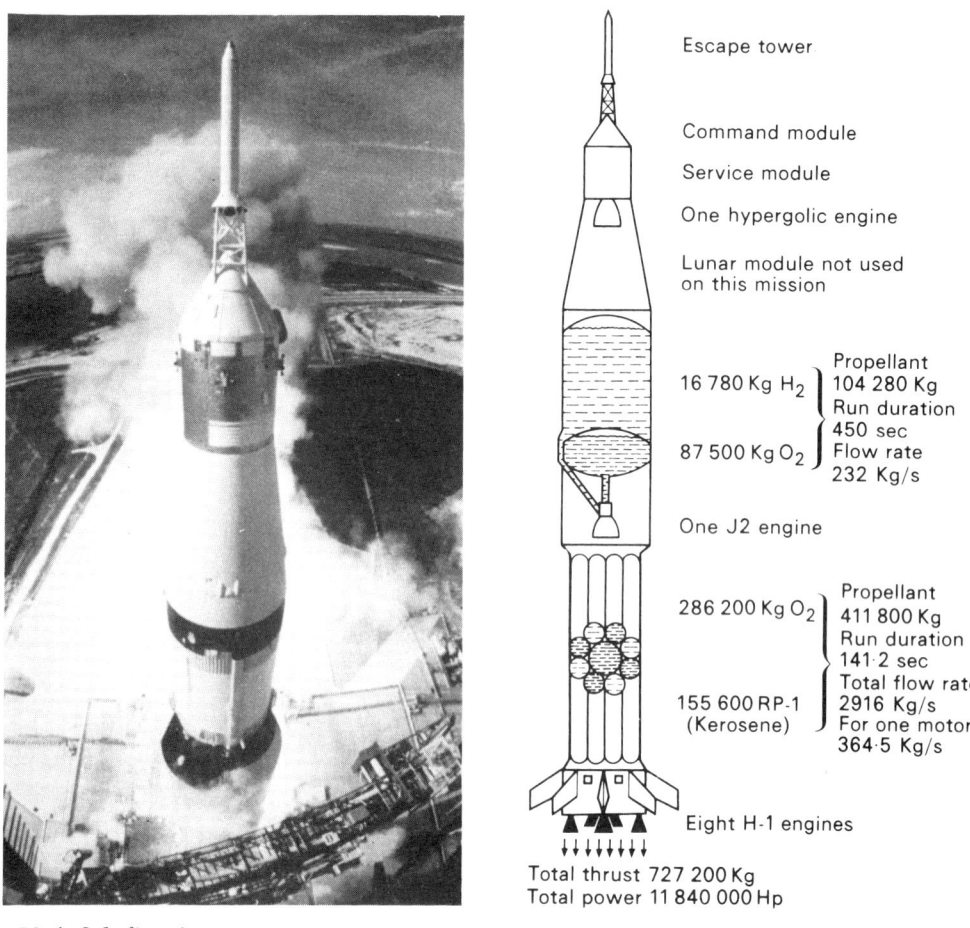

Escape tower

Command module

Service module

One hypergolic engine

Lunar module not used
on this mission

16 780 Kg H$_2$ } Propellant
104 280 Kg
Run duration
450 sec
87 500 Kg O$_2$ } Flow rate
232 Kg/s

One J2 engine

286 200 Kg O$_2$ } Propellant
411 800 Kg
Run duration
141·2 sec
Total flow rate
155 600 RP-1   2916 Kg/s
(Kerosene)   For one motor:
364·5 Kg/s

Eight H-1 engines

Total thrust 727 200 Kg
Total power 11 840 000 Hp

FIG 1.2 *left*   Apollo-11 moon mission. *(Photograph by courtesy of
the United States Information Service.)*

right   A sectional view of a rocket. *(From Thermodynamics
Atlas 2 by Ivo Kolin.)*

Within the engine, processes occur that make the energy suitable for
transfer out of the engine by work which, in its simplest form, can be represented
by the rotation of a shaft. This energy W is shown leaving the heat engine E in
Fig. 1.5.

FIG 1. 3 (a)   A passenger ship, QE2.

(b)   **A boiler in QE2.** *(Photographs by courtesy of the Cunard Steamship Company Limited.)*

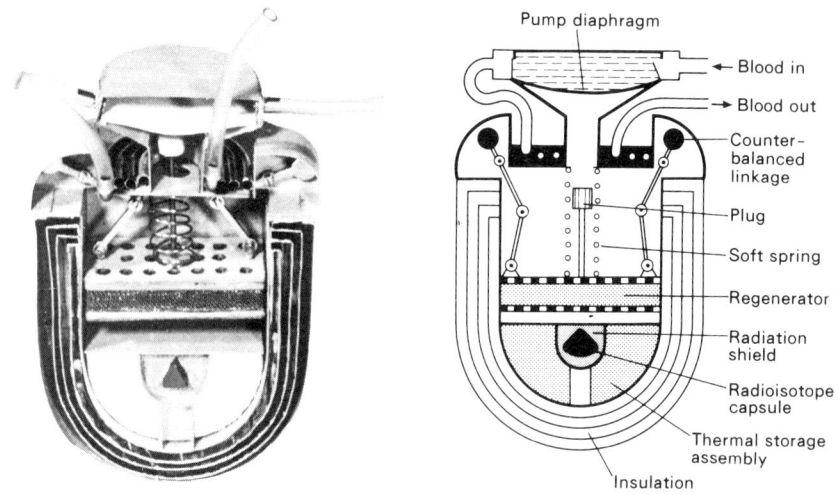

FIG 1.4 *left*   A model of a nuclear-powered artificial heart.

right   A sectional view of the heart. *(From Thermodynamics Atlas 2 by Ivo Kolin.)*

1.4   The thermodynamic heat engine

The heat-engine represented in the thermodynamic sense by a box in Fig. 1.5 consists of all the components required to convert heat energy Q_1 into work energy W. This energy conversion is achieved by a fluid undergoing a series of changes when passing through the various components of the heat-engine. A complete heat-engine usually comprises a heater, turbine (expander), cooler and compressor (or pump) although the functions of some of these may be combined in practice in a single engineering component. Fig. 1.6 shows the arrangement of components in a simple heat-engine. All the components are within the box of Fig. 1.5.

If the fluid were water the heater would be called a boiler and would then be like a kettle. Unlike a kettle a boiler is continuously operating, being fed with cold water and giving off steam continuously. That is why we normally draw diagrammatically a heater in the form shown in Fig. 1.6(a). The fluid is entering from the feed pipe marked 1 in the figure, is supplied with energy at the rate $\dot{Q}_1$ and is leaving through the pipe marked 2. The energy may be supplied from any combustion process, nuclear reaction, or electric heating element.

The turbine, Fig. 1.6(b), is not like a kettle but is like a windmill. The fluid, wind, strikes the sails of the windmill and causes the shaft to which the sails are attached to turn, and in turning, to transfer energy by work. In this case

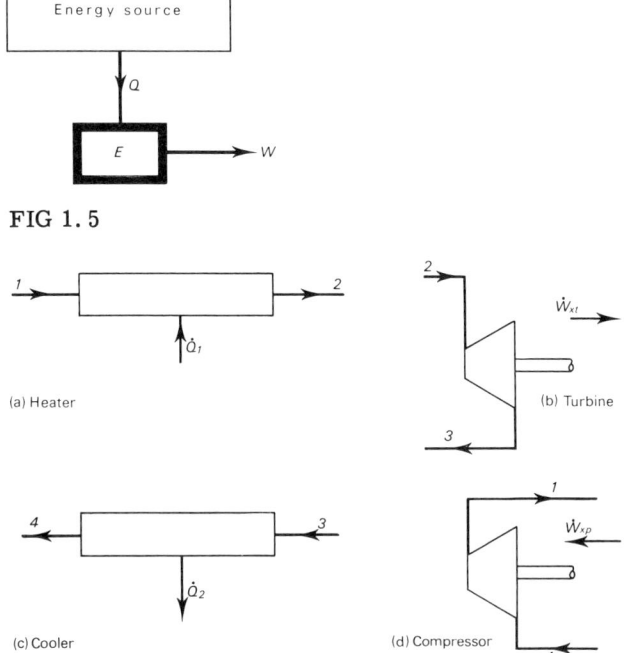

FIG 1.5

FIG 1.6   Diagrammatic Representations of Four Processes

the working fluid loses energy as it goes from 2 to 3, and this is transferred by work out of the component being considered.

In the two components already considered note that two types of working fluid have been mentioned. In the first—the kettle—the fluid is present both as a liquid and a vapour whereas in the second—the windmill—the fluid is only present as a gas. This is important because there are basically two classes of power plants, vapour power plants and gas power plants.

Let us complete the cycle if the working fluid were water. The cooler would be called a condenser. This would be similar to a cold window in a stuffy room on which the moisture would condense in drops. In a condenser, metal tubes take the place of the glass in the window over which a continuous flow of steam is maintained. We draw a condenser diagrammatically in the form shown in Fig. 1.6(c). The steam from the turbine enters the condenser at point 3 and loses energy by heat to the coolant passing through the condenser tubes and the water emerges from the condenser as a liquid at point 4.

Finally, there is the compressor, or feed pump as it is called in a steam plant because it feeds water to the boiler. The compressor takes the fluid from the cooler through the pipe marked 4 in Fig. 1. 6(d), gives it additional energy by compressing it to a higher pressure and delivers it to the boiler through the pipe marked 1. The characteristic shape of the turbine and compressor shown in Fig. 1. 6(b) and (d) show changing cross-sections because as the pressure of a fluid increases or decreases so does its volume increase or decrease.

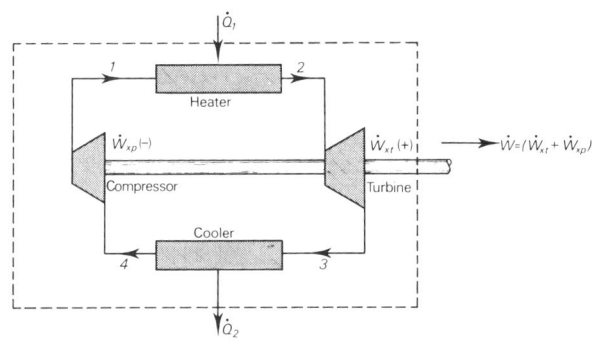

FIG 1. 7   A simple Heat Engine

1. 5   Thermodynamic cycle

When the heater, turbine, cooler and compressor are joined together they form a very simple heat engine as shown in Fig. 1. 7, a thermodynamic cycle of four components is formed. Two principles which apply to the heat engine will emerge during the course of the book; the first is that energy must be conserved and the second is that to produce work energy, W, from heat energy Q_1, there must be a second heat transfer, Q_2.

What has been written in section 1. 4 about the components and the thermodynamic cycle describes the fundamental components for a steam and for a gas power plant. The same principles apply to the analysis of a reciprocating internal combustion or reciprocating steam engine. These principles also apply to the cycle operating in reverse when the cycle operates as a refrigerator and as a heat pump.

In this book we consider processes involved in the components making up a heat engine and study the properties of different working fluids in undergoing these processes.

2 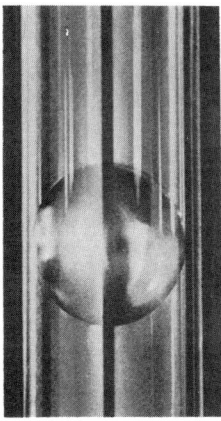 *A system*

A system, its properties, and the actions of heat and work on its boundary are described and related to the first law. Cyclic and non-cyclic processes are introduced.

2.1 A system

In the first chapter mankind's desire to invent a magic box that would do work was described. So that we may discuss in more detail the problems of designing such a box we must define certain terms which we will be using.

Let us call the thing under consideration, in this case the magic box, the system, and the rest of the Universe **THE SURROUNDINGS.** No mass may be transferred either way across the boundary of a system although the shape of the boundary may change. The system shown in Fig. 2.1(a) may be the same as that shown in Fig. 2.1(b) in spite of the difference in their shapes. If it is the same system the mass within the boundary in Fig. 2.1(a) is the same mass, no more nor less, than that within the boundary in Fig. 2.1(b).

A system has properties of which various forms of energy are of greatest importance to the engineer. A **SYSTEM** therefore is an unchanging quantity of matter in which energy can be stored. Energy may flow in both directions between the system and the surroundings.

Systems (Q and A)

Q. Within our definition of a system can the following be regarded as systems?

    (a) A block of steel
    (b) A sponge
    (c) A pool of petrol

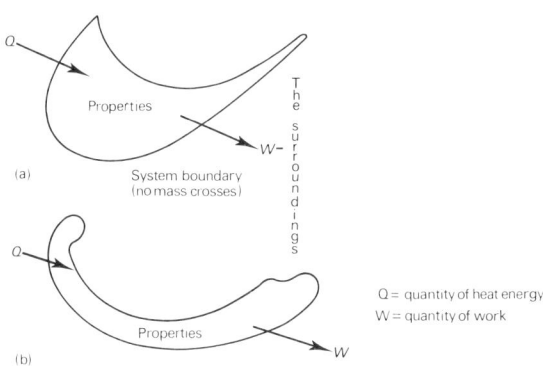

FIG 2.1   The same system changed in shape

A.   The difficulty in answering this question lies in the initial choice of the system boundary. This having been done, (a) and (b) but not (c) can be considered systems. Some of the difficulties in choosing the boundary definitions are given below.

(a)   The boundary of a block of steel is relatively easy to define, but it would not have been so easy if the block were wood, which is porous.
(b)   Because a sponge, like a block of wood, is continually taking in and giving out air and moisture the boundary is not easily defined, although it can be done. The fibres of the sponge must clearly be within and remain in the system, but the spaces through which the air and moisture pass must be excluded. That is to say the boundary must, in this case, be drawn around the solid matter if the sponge only is to be the system. A system can consist of a mixture of ingredients, however. If the boundary were drawn around the sponge and a fixed mass of air and moisture with it, the system within the boundary would consist of the sponge, air and moisture.
(c)   The word 'pool' implies a liquid bounded by a liquid-vapour interface where evaporation and condensation are continuous. This is how liquids and vapours behave. If the 'pool' of petrol is wholly liquid as is commonly supposed, such changes of phase are not admissible. Therefore a pool of petrol, as ordinarily understood, can never be a system. If the boundary were drawn to include the vapour then it is a system but can hardly be called a pool.

2.2   Heat and work

Energy may be transferred into and out of a system because of a temperature difference between the system and its surroundings. This mode of transfer of energy we call heat or heat transfer. **HEAT** is defined as the action of energy transfer that occurs on the boundary of a system by reason of temperature differences between it and its surroundings.

Energy may also be transferred into or out of a system by interaction between a system and its surroundings unconnected with temperature differences. For instance, energy may be transferred by an electric current passing across the system's boundary. Energy may be transferred by a force being applied to the boundary of a system and changing the system's shape, such as in the case of metal being forged by a hammer. Energy will be transferred when forces are applied to overcome friction where two solid systems are rubbed together, and the increase of stored energy in the solid systems will be seen as a rise in their temperatures. Such modes of transfer we call work or work transfer. In any of these modes of transfer, the electric current or the moving force, the energy transferred could have been stored by using the energy to raise a mass against the gravitational pull of the Earth whereas the other means of energy transfer, heat, could not have been used solely to do this (see Chapter 5). **WORK** is defined as the action that occurs on the boundary of a system when energy is transferred in such a way that the energy could have been used solely to raise a mass against the force of gravity.

Sign conventions

When the surroundings are at a higher temperature than the system energy is transferred by heat, and this heat transfer of energy is defined as positive because energy is transferred *inwards* from the surroundings to the system. When the system does work on its surroundings this work transfer of energy is defined as positive because energy is transferred *outwards* from the system.

So heat is positive when operating into a system and work is positive when operating out of a system. The reason given for apparently conflicting signs to the two modes of energy transfer is quite simple. Many systems, but not all, take in energy Q by heat and deliver energy W by work. Such a system is called a heat engine if it operates in a cycle—that is if it returns to its initial condition at the end of the transfers of Q and W. This particular case of a system operating in a cycle is basic to an engineer's application of thermodynamics to many practical examples. If the sign convention given above is used for a system that undergoes a cycle it is commonsense that the transfers of Q and W must balance because the system returns to its initial state and so,

$$Q - W = 0 \qquad\qquad (2.1)$$

2.3 Properties

The condition of a system at a given time is called the **State** of the system, and the state of the system is described by the **properties** of the system. The number of properties required to define the state of the system depends on the complexity of the system. While its properties are changing a system is changing state and is said to be undergoing a **Process**. While the properties are changing during a process, the system goes from one state to another passing through an infinite number of intermediate states as shown in Fig. 2. 2 and the line joining these intermediate states is called the **Path** of the process. At the end of a non-cyclic process such as that shown in Figs. 2. 2 and 2. 3(b) one or more of the properties will be different at the end of the process from what it was at the beginning.

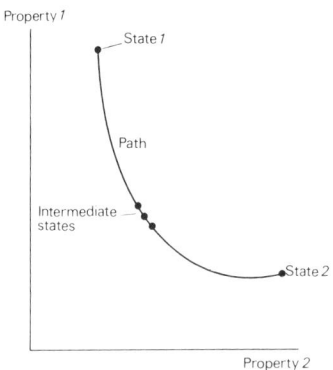

FIG 2. 2   The path of a process

An important point to remember about properties is that they are related to the state only of the system, not of the path taken to reach that state. For instance (see Fig. 2. 3(a)), if the state of a system changes from state 1 to state 2 it does not matter, so far as the properties at state 2 are concerned, whether the change of state from 1 to 2 was made along path A or path B. From a concept's dependence or otherwise on the path taken one knows if one is dealing with a property or not. This can be illustrated by the height above sea-level of a man climbing a mountain. The man's height above sea-level at any place is a property and is independent of the path he took to reach that spot. From whatever direction he comes, and no matter how long he took to get there, his height above sea-level will always be the same when he reaches that place, and therefore we recognise this as one of his properties. It is also a property of that part of the mountain.

Properties (Q and A)

Q.   Could the following concepts be regarded as properties of the systems stated? In each case work and heat transfer may occur to change the system from one state to another.

System	Concept
(a) A bucket containing water	temperature
(b) An electric battery	stored charge
(c) Incandescent electric fire	radiant energy
(d) Automobile engine	power output
(e) A weather balloon	pressure

A. Ask yourself 'does the concept help to define the state of the system independently of the many possible routes it took to reach that state?'. If so, it is a property; if not, it is not a property.

> (a), (b) and (e) Each of the concepts help to define the state of the system independent of the route taken to reach that state. Hence they are properties.
> (c) This concept is energy that has left the system and therefore the answer to the question is 'no'. It is not a property.
> (d) This concept is energy already transferred out of the system. The answer is 'no'. It is not a property.

2.4 Processes—cyclic and non-cyclic

Earlier the word non-cyclic was used in describing a process that led to a change of state. Consider a system in a certain state, and assume it undergoes a process of such a kind that the system's state at the end of the process is the same as it was at the beginning. In Fig. 2.3(a) one possible process, 1A2B1, is a cyclic process. We know this because the system is in the same state, state 1, at the end as at the beginning of the process. Such a process is said to be **cyclic**.

In Fig. 2.3(b) is shown a possible process from state 1 to state 4, 1C3D4. We know this is a non-cyclic process because the system is in a different state, state 4, at the end from that at the beginning, state 1. Such a process is said to be **non-cyclic.**

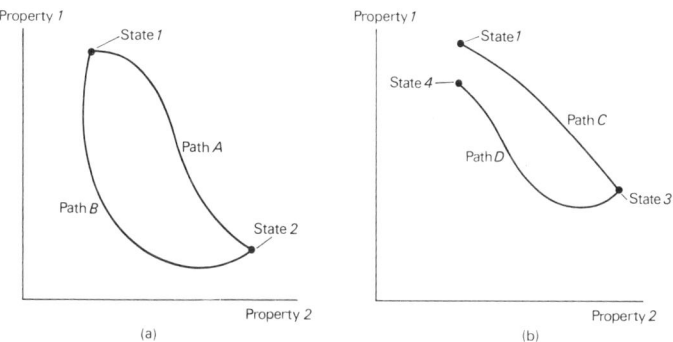

FIG 2.3   Paths between states

Processes (Q and A)

Q. Can any of the following systems be described as undergoing a cyclic process?

> (a) A *block of metal* is carried into a tall building, taken to the top and dropped from a window, landing on the spot where it started.

(b) A *man* walks into a block of flats, takes a lift to the twentieth floor, walks to the window and jumps out—landing on the spot from which he started.

(c) A *ship* powered by a nuclear reactor moves from port A to port B and then returns to port A.

(d) A gold *ring* is removed from its mother's finger by an infant, who swallows the ring! Later, the ring is recovered and replaced on its owner's finger.

(e) A *sailing dinghy* leaves the basin for a day's sailing. Later it returns to the same basin.

A. Ask yourself the question; 'Have all the properties of the system remained unchanged?'

(a) (i) If its shape remains unchanged it has undergone a cyclic process.

(ii) It it is permanently deformed it has undergone a change of state, and the process was therefore not cyclic.

(b) No—the man will be dead or at best badly bruised.

(c) No—the ship will have used some of its fuel.

(d) and (e)—Yes.

2.5 The first law of thermodynamics

The first law of thermodynamics tells us that energy must be conserved, or in a cyclic process the energy transferred by heat equals the energy transferred by work. Mathematically this is written

$$\oint dQ - \oint dW = 0 \tag{2.2}$$

or, in the case of the process 1A2B1 shown in Fig. 2.3,

$$Q - W = 0 \tag{2.3}$$

This is equation (2.2) integrated around the cycle and is the same as equation (2.1)

If the process under consideration were not cyclic; if for instance heat energy Q were greater than the work energy W, some of the properties of the system, see Fig. 2.1, would be changed and the system would undergo a change of state. The difference between these transferred energies, Q and W, has changed the system's store of energy. Figure 2.4, in fact, depicts more precisely such a system, and equation (2.3) derived from the first law, would, if applied to this system undergoing a non-cyclic process, be

$$Q - W = \Delta E \tag{2.4}$$

ΔE being the change of energy in the system associated with its change of state.

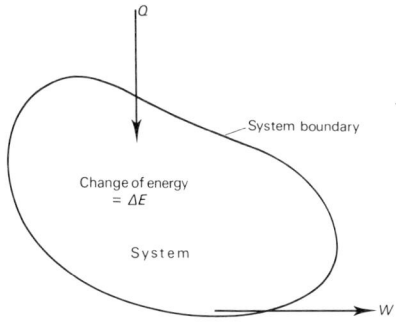

FIG 2.4   A system

Heat, work and energy (Q and A)

Q. A lawnmover is being used to cut grass. With the lawnmower as the system consider a process that comprises the cutting of grass; state the sign of the heat, work and changes of energy in the system and in its surroundings during the process if the mower is driven

    (a) entirely by a man
    (b) by a small petrol-engine (fuel is carried as part of the mower hence it must be considered as part of the system)
    (c) by a mains-operated electric motor.

A. It will be observed by the reader that every answer given below includes a negative heat transfer. This is due to a part of the work transfer causing rubbing between the moving and the stationary blades, shown by a rise of temperature inside the mower. This rise of temperature causes the negative heat transfer.

	System	Surroundings
Q	−	+
W	−	+
ΔE	+	−

    With regard to the W, the man does work on the mower which transmits this work energy W_m to the common surface between the mower and the ground. There the energy splits into two divisions. One division is transmitted out of the mower by work in two parts,
(1) the crushing of the ground where the wheels have passed, and
(2) the cutting of the grass.
One hopes that the energy expended on the second of these is the greatest part of the original work energy W_m. The other division remains in the mower also in two parts,

(3) the rubbing of the moving on the stationary blades, and
(4) the overcoming of friction in the bearing.

These last two parts of the original work energy W_m increases the energy stored
in the mower by an amount ΔE causing a rise of temperature and the subsequent
negative heat transfer mentioned above. While the mower is moving steadily the
whole of the work energy W_m goes into (1), (2), (3) and (4) above, but when the
mower is restarted after a stop some goes into kinetic energy.

	(b) System	Surroundings
Q	−	+
W	+	−
ΔE	−	+

This case is the same as (a) except that it is expenditure of internal energy of
fuel that makes the mower do work on the grass.

	(c) System	Surroundings
Q	−	+
W	−	+
ΔE	+	−

This is the same case as (a) except that electric power from the mains does work
on the grass instead of the man.

2.6 Summary

A system is an unchanging quantity of matter, in which energy may
be stored.

Heat is the action that occurs on the boundary of a system when
energy is transferred by reason of temperature differences between it and its
surroundings.

Work is the action that occurs on the boundary of a system when
energy is transferred in such a way that the energy could have been used solely
to do work against gravity (i.e. raise a mass).

The state of a system is a description of as many properties as are
required to define the system completely. When one or more properties change
in a system, the system is said to have changed its state and undergone a non-cyclic
process in the course of which it passes through a series of intermediate states
called the path of the process. The properties of a system are a part of its state
and are independent of the path by which this state was reached. Temperature,
pressure, volume, height, velocity and composition are some of the properties of
a system. When any of these change the stored energy E changes.

The change of stored energy ΔE of a system is related to the heat
energy Q and work energy W between the system and its surroundings according
to the first law of thermodynamics which states that

$$Q - W = \Delta E \tag{2.4}$$

2.7 Questions for the reader

Q.1. A house uses electric storage heaters that operate from an off-peak supply. Consider the house with its storage heaters to be the system and state the signs of heat, work and energy changes in the following circumstances

(a) The temperature outside the house is lower than inside but the electricity is turned off.
(b) Starting from situation (a) the power has very recently been switched on.
(c) The power has been on for a very long time.

A.1.

	Q	W	ΔE
(a)	−	0	−
(b)	−	−	+
(c)	−	−	0

Q.2. For the systems given below, undergoing the processes stated, are the heat and work actions and changes of stored energy positive, negative or zero, in each case?

(a) *A steam generator,* which is in use, is shut down, all power is cut off and valves closed. It is left to cool.

$$(-,0,-)$$

(b) *One kg of gas flows into an evacuated steel bottle* from a high-pressure main There is no heat transfer.

$$(0,-,+)$$

(c) *One kg of gas* at high temperature expands in a cylinder as the piston moves slowly outward. The cylinder, the piston and the rest of the surroundings are at a lower temperature.

$$(-,+,-)$$

(d) *One kg of an ice and water mixture* is contained in a closed rigid vessel which it completely fills. A flame is applied to the outside walls of the vessel until some, but not all, of the ice has melted.

$$(+,0,+)$$

(e) What would the answer to (d) be if the walls of the container had been elastic

$$(+, -, +)$$

In this answer W is negative because 1 kg of water has a smaller volume than the same mass of ice at the same temperature. There is therefore a reduction in volume of the system and the surroundings do work on it.

Q. 3.   For the systems given below, undergoing the processes stated, are the heat and work actions and change in stored energy positive, negative or zero in each case. Give brief justifications for your answers.

(a) *An electric battery*—it is being used to drive a small motor.

$$(-, +, -)$$

The chemical change will decrease the battery's stored chemical energy and increase its temperature. Work is being done by the current.

(b) *An automobile engine with its fuel tanks*—together are used to drive a car.

$$(-, +, -)$$

The engine will be cooled probably by air and water. Work is being done by the crankshaft.

(c) *A gas-fired water radiator heating system*—it has been working at a steady temperature for a long time.

$$(-, -, 0)$$

A very small amount more energy will be lost from the radiators than will be gained from the gas flame because an electric motor will be using a small amount of energy in driving the circulating pump, transferring extra energy by work into the system. ΔE will be zero because the system has run a long time and the energy remains constant.

3

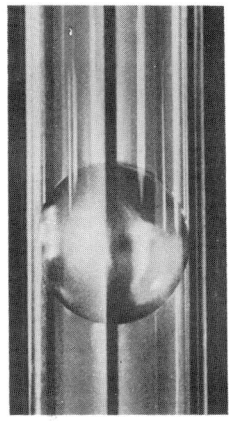

Heat, work and energy

In the last chapter the first law of thermodynamics was stated in the form

$$Q - W = \Delta E$$

Now the actions of work and heat and the consequent changes of stored energy, ΔE, are considered in more detail. Changes of stored energy are split into changes in internal, kinetic and gravitational energies.

3.1 The relationship between heat and work

There has been much misunderstanding about the relationship between heat and work. It has arisen mainly because of the historical ideas about heat, many of which are still with us today in our colloquial use of the word.

In every day usage heat means the action of transferring energy and also the energy that is transferred. An additional confusion is that the work heat is used for stored energy whether it was transferred by work or heat to the system. In the thermodynamic sense work and heat are actions at a system's boundary.

When work is done *by* a system, defined as positive word in section 2.2, energy is lost from the system and gained by the surroundings. When this action of transferring energy, called work or sometimes work transfer, is finished the energy has been transferred and the work is over and done with. If you push an object up a slope it is quite clear that you are doing work on it and that, when you stop pushing, the work is finished. Also it can be readily understood that as a result of this work gravitational energy has been stored in the object by virtue of its higher position on the slope. Few would insist that work has been stored.

Similarly, we are familiar with the fact that if we do work on a fluid by stirring, see Fig. 3.1, a rise of temperature occurs. In this case work has been done on the liquid system and the property that we call energy, E, has been increased. No heat, in the thermodynamic sense, has occurred in this operation.

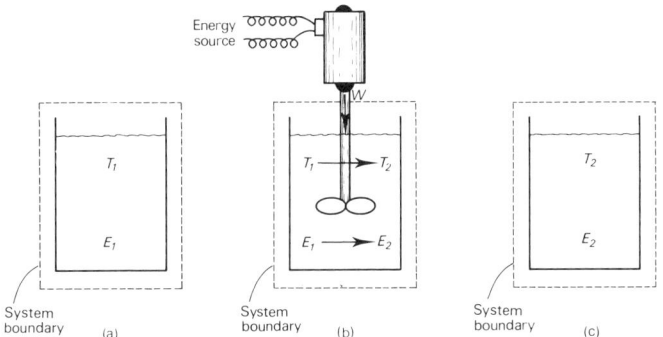

FIG 3.1   Work causing an increase in a system's energy

Because heat and work are both modes of energy transfer between the system and its surroundings the same points can be made about heat as have been made about work. Heat, sometimes called heat transfer, is also an action on the boundary of a system and it ceases to exist after the transfer of energy ceases. If heat is done *on* a system because the temperature of the surroundings is greater than that of the system, defined as positive heat in section 2.2, additional energy ΔE is transferred to the system. If, for example, energy is transferred by heat from the ground to a pocket of air the temperature of the air will increase due to an increase of energy in the system. The air pocket will then rise and while gaining height it may carry with it a bird or even a glider to a greater height. So work has been done on the glider. If the pocket of air is the system under consideration it has taken in energy by heat from the ground and given out energy by work in raising the glider to a greater height.

We are also familiar with the fact that if we transfer energy to a fluid by heat, as shown in Fig. 3.2, a rise in temperature occurs. There is no clear line of demarcation between a solid and a liquid but we call a substance a **Liquid** when it appears to offer no resistance to a slow change of shape. In the case of Fig. 3.2, heat has transferred energy to the fluid system and so we say that the energy has been increased from E_1 to E_2—just the same as in the case of work—and the increase in energy is apparent from a rise in temperature of the fluid. In the case illustrated in Fig. 3.2, we must *not* say there has been a gain of heat by the fluid because, as in the case of the stirrer illustrated in Fig. 3.1, once the energy source has been removed heat ceases as surely as work ceases and we need have no knowledge of how the energy was increased.

Heat and work (Q and A)

Q.   If a man, wearing a thermally insulating suit, pushes for a long time against a rigid wall he will probably die primarily from what is called 'heat-stroke'. Considering the man and his sweat as the system, but neglecting the effect of his

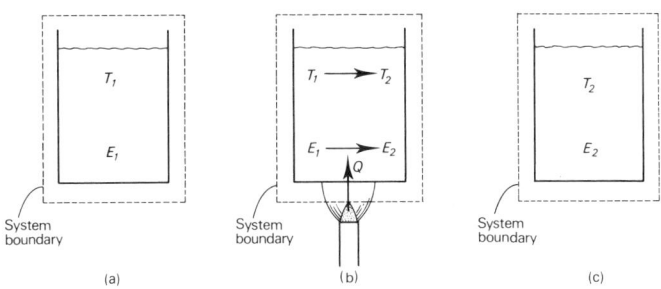

FIG 3.2   Heat causing an increase in a system's energy

breathing which carries only a very small quantity of energy, has there been (a) any work? (b) any heat? What changes in energy have there been?

A.   (a)   There has been no work because the wall is rigid.
(b)   There has been no heat because the man is thermally insulated. The main change of energy has occurred within the system; the chemical energy in the man's muscles has changed to another form of stored energy. This is a change that occurs in an animal's body to sustain a force such as the useless force that the man is applying to the rigid wall. The change from chemical energy causes a rise in the man's temperature. However, $\Delta E = 0$ and the total energy of the system has remained constant although some of it has changed in form.

3.2 Heat colloquially and in thermodynamics

Strict adherence to definitions is often necessary in engineering so that we all understand precisely the same meaning of a word or phrase. Colloquially heat is given any one of the following meanings,

(a) The action of transferring energy by reason of a temperature difference, as when we say the flame heats the saucepan'.
(b) Energy that is transferred whether by work (Fig. 3.1) or by heat (Fig. 3.2), as when we say 'if I rub two blocks together heat is generated' or 'heat from the flame makes the saucepan hot'.
(c) Energy that is stored in a system, as when we say 'it is heat that gives the water its high temperature'.

In thermodynamics the engineer uses only meaning (a), because this employs 'heat' as denoting an action whereas the others use 'heat' as denoting an energy. That is why we define **HEAT** as the action of energy transfer that occurs on the boundary of a system by reason of temperature differences between it and its surroundings.

3.3 Work in thermodynamics, mechanics and electricity

The engineer is rarely concerned with one engineering discipline. Usually he must be ready to cross the disciplinary boundaries and use words like heat and work in situations in which thermodynamical, mechanical and electrical problems are intermingled. Not only, therefore, must his meaning for the word be closely defined as in the case of the word 'heat' but the definition must hold in a varieity of disciplines. For the word 'work' the reader will probably be familiar with the applied mechanical definition that states that work energy is **force x distance** which, while being correct so far as it goes, does not include electrical work.

Work in thermodynamics is defined in such a way that electrical work is included. It is defined as the action that occurs on the boundary of a system by which energy is transferred in such a way that it could have been used solely to raise a mass against the force of gravity. In Fig. 3.3(a) work is being done by a

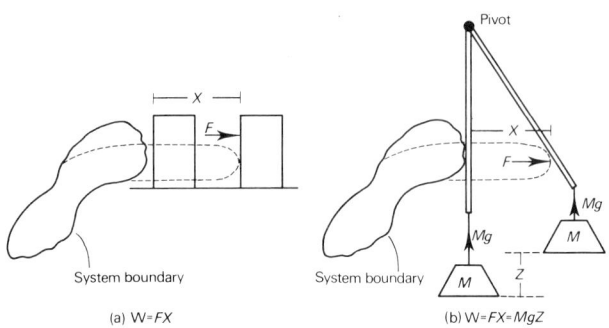

FIG 3.3   Work according to both the applied mechanical and the thermodynamical definitions

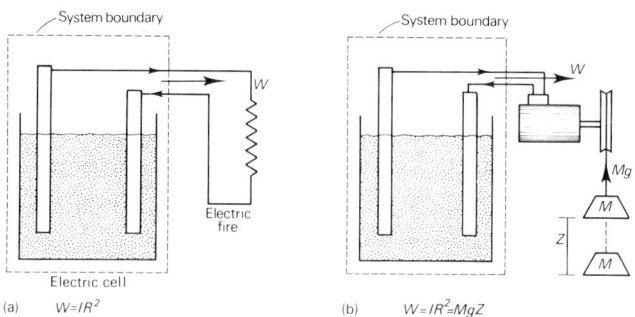

FIG 3.4   Work according to only the thermodynamical definition

system in that it is pushing a block a distance X with a force F. The applied mechanical definition covers such a case and evaluates the work energy W (that is the energy being transferred by work) as F × X. The thermodynamical definition also covers the case in that the energy could have been used solely to raise a mass as in Fig. 3. 3(b). However, the mechanical definition does not cover the case in Fig. 3. 4(a) whereas the thermodynamical definition does. The thermodynamical definition covers the case of Fig. 3. 4(a) because the action on the boundary—the electrical potential and current, operating over a period of time—could have been used solely to raise a weight, as in Fig. 3. 4(b).

3. 4   Energy

When energy Q is transferred by heat to a system and energy W is transferred by work from a system, Q and W not being equal, the stored energy E in the system will be changed by an amount ΔE, and equation (2. 4) stated that, for such a system,

$$Q - W = \Delta E$$

or, in words, the change ΔE of energy stored in the system is the difference between the total energy transferred by heat, Q, and the total energy transferred by work, W.

The energy E in a system is the sum of a number of forms of energy, some of which are familiar and some are not. Energy can be stored in the form of internal energy U, kinetic energy K, gravitational energy Z, and other forms of stored energy associated with surface tension, electric and magnetic fields, etc. In classical engineering, because of the nature of the problems considered, forms of energy other than internal, kinetic and gravitational are negligible, just as in classical engineering Newton's second law is sufficient without using Einstein's relativity theory. Therefore in most engineering thermodynamical problems,

$$E = U + K + Z + \ldots \tag{3.1}$$

the other terms being negligible.

Then equation (2. 4) can be written more fully as,

$$Q - W = \Delta U + \Delta K + \Delta Z \tag{3.2}$$

3. 5   Internal energy

The more one studies the three forms of energy U, K and Z, the more apparent it is that there is not a clear-cut division between them. What is called internal energy U has four component forms;

(a)   Energy in the random movements of the molecules relative to the total system.

(b) Potential energy due to forces between molecules.
(c) Potential energy due to forces between atoms within a molecule.
(d) Potential energy due to forces between elementary particles
within an atom.

Even if the total internal energy were to remain constant the energy
may change between any of the four types. For instance, in the problem discussed
at the end of section 3.1 the internal energy in the man is changing from chemical
energy of form (c) to form (a).

Usually we are discussing fluids that are chemically stable and in
these cases the change of energy of a fluid of mass M when its temperature is
changed from T_1 to T_2 is found by using a property known as specific internal
energy per degree. Under the International System of units the term for this property is specific heat capacity but, because this is an improper use of the word
heat, we will use the more descriptive term specific internal energy per degree.
This allows the change of internal energy in the case of a chemically stable fluid
to be expressed in the equation

$$\Delta U = M\ c_V\ (T_2 - T_1) \tag{3.3}$$

where c_V is called the specific internal energy per degree. The units of c_V are
energy per unit mass per degree. Specific energies are defined more fully in
Chapter 18.

Internal energy (Q and A)

This example of a balloon is used throughout this chapter, more information being
given as it becomes necessary.

Q. The air in a hot-air balloon has a c_V of 1 kJ/kg. Ten kg of the air is heated
from 20 to 80°C during the first minute after being unmoored. By how much has
its internal energy changed?

A. Substituting in equation (3.3), M = 10 kg. c_V = 1 kJ/kg K.

$$T_2 - T_1 = (353 - 293)\ K$$

then the change of internal energy U is

$$\Delta U = 10 \times 1 \times 60$$
$$= 600\ kJ$$

3.6   Kinetic energy

The kinetic energy K of fluid of mass M is the energy that the fluid
has by virture of its motion relative to some datum such as the Earth which for
convenience is considered stationary. It is equal in magnitude to the work energy
W that it would do against a force F if, in the absence of any heat transfer, the

force changed the velocity of the mass M from V to zero. The following statement can be made

$$K = W_{V \to 0}$$
$$= - F \times \text{ distance through which F is applied}$$

(the negative sign is appropriate because the system is doing work against the force F)

$$= \int_V^0 FV \, dt$$

if F is variable as well as V, and t is time,

$$K = - \int_V^0 \frac{M}{dt} \, dV \, V \, dt$$

$$= - \left[M \frac{V^2}{2} \right]_V^0$$

$$= + \frac{M V^2}{2}$$

Therefore $K = \frac{1}{2} MV^2$, and a change ΔK of kinetic energy stored in the system when its velocity is changed from V_1 to V_2 is given by

$$\frac{\Delta K}{1-2} = \int_1^2 MV \, dV$$

$$= \left[M \frac{V^2}{2} \right]_1^2$$

$$= \frac{1}{2} M \, (V^2 - V_1^2) \tag{3.4}$$

Kinetic energy (Q and A)

Q. During the first minute the velocity of the balloon increases to 10 m/s. What is its increase of kinetic energy?

A. Substituting in equation (3.4), M = 10 kg, $V_1 = 0$ and $V_2 = 10$ m/s, the change of kinetic energy is

$$\Delta K = \frac{10 \, (10^2 - 0)}{2}$$

$$= 500 \text{ J or } 0 \cdot 5 \text{ kJ}$$

3.7   Gravitational energy

A system has gravitational energy by virtue of its mass and the strength of the gravitational field in which it is sited. If the mass of the system is M in a gravitational field of strength g then from Newton's laws its weight—that is to say the force that is pulling the system towards the centre of the gravitational field—is Mg. If M is in kg and g in m/s² then the force is in newtons. On Earth we live in a relatively powerful gravitational field and hence this form of stored energy often cannot be neglected. If work is done on a system of mass M, initially at a distance $\bar{z}_1$ from the centre of a uniform gravitational field, to cause the system to move to a new position $\bar{z}_2$ from the centre of the field, the change of gravitational energy Z is

$$\Delta Z = \text{(force due to gravity) x (distance moved in direction of force)}$$
$$= Mg\,(\bar{z}_2 - \bar{z}_1) \tag{3.5}$$

Gravitational energy (Q and A)

Q. If the hot-air balloon has been released at about sea-level and rises, in the first minute, to a height of 40 m, what has been the increase in its gravitational energy? The strength of the gravitational field is 9·8 m/s².
A. Substituting in equation (3.5)M = 10 kg, g = 9·8 m/s², $\bar{z}_2 = 40$ m, $\bar{z}_1 = 0$, the change of gravitation energy Z is

$$\Delta Z = 10 \times 9\cdot8\,(40 - 0)$$
$$= 3920 \text{ J or } 3\cdot92 \text{ kJ}$$

3.8   What thermodynamics is about

Equation (3.2) gives a relationship between Q the energy transferred into a system by heat, W the energy transferred out by work, and the resulting changes in the stored energy—internal energy ΔU, kinetic energy ΔZ—in the form

$$Q - W = \Delta U + \Delta K + \Delta Z \tag{3.2}$$

This is what thermodynamics is about. **THERMODYNAMICS** is a study of the energy transfers between a system and its surroundings, and the resulting changes in the various types of stored energy within the system.

Relationship between Q, W, ΔU, ΔK and ΔZ (Q and A)

Q. During the first minute after launching, the balloon drags a chain a distance 10 m along the ground against a frictional force of 60 N. Assume that any work done by the balloon on the atmosphere is negligible. With all the previous information find the energy transfer by heat to the balloon. Omit any consideration of the density of air varying.
A. W = 60 × 10 = 600 N m = 0·6 kJ

$$\Delta U = 600 \text{ kJ (section 3·5)}$$

ΔK = 0·5 kJ (section 3·6)

ΔZ = 3·9 kJ (section 3·7)

Substituting these values in equation (3. 2),

$$Q = \Delta U + \Delta K + \Delta Z + W$$
$$= 600 + 0·5 + 3·9 + 0·6$$
$$= 605 \text{ kJ}$$

3.9   Summary

The definition of work was examined and expanded to include not only the mechanical definition but also the cases in which electrical work exists. The definition of heat was examined by considering its colloquial uses and restricting that definition to a single specific meaning useful in thermodynamics. The relationship between energy transfers and changes in different types of stored energy were considered leading us to the equation

$$Q - W = \Delta U + \Delta K + \Delta Z + \text{(other terms which may normally be neglected)}$$

3.10   Questions for the reader

Q. 1. A lift cage with its full complement of passengers has a total mass of 325 kg. Calculate the energy required to raise the cage and its load to the top of a lift shaft 50 m high.

[159 kJ]

Q. 2. If, when the lift cage is still fully loaded and at the top of the shaft, the braking mechanism failed what, assuming friction is negligible, would be the kinetic energy of the cage when it struck the bottom of the shaft?

[159 kJ]

Q. 3. The mechanical efficiency of a bicycle is the ratio

$$\frac{\text{energy usefully employed}}{\text{energy expended by rider}} \times 100$$

If the efficiency is 80 per cent, has the 20 per cent gone to internal, kinetic or gravitational energy?

[By work through friction to internal energy, then by heat transfer.]

Q. 4. A man whose mass is 70 kg is a keen cyclist and wishes to calculate the energy he expends in carrying out certain manoeuvres. The first of these is climbing a hill of 300 m vertical height at constant velocity on a bicycle of 10 kg mass.

What energy has he expended if the bicycle's mechanical efficiency is 90 per cent?

[261 kJ]

Q. 5. In the second manoeuvre the cyclist of question 4 starts from rest and accelerates to 10 m/s on level ground. How much energy has he expended?

[4·44 kJ]

Q. 6. In the third manoeuvre the cyclist mounts his cycle at the top of the hill and freewheels down the hill until a velocity of 10 m/s is attained. After this he uses his brakes to keep the velocity at 10 m/s until he reaches the bottom of the hill when the brakes are released. Calculate the energy transferred by heat from the brakes and wheel rim by the time the brakes have returned to normal temperature

[231 kJ]

Q. 7 A cannon-royal of 3 600 kg mass fires a ball of 28 kg mass. If the cannon initially moves backwards with a velocity of 1 m/s use the principle of conservatic of momentum to calculate the initial kinetic energy of the ball.

[231 kJ]

Q. 8. In the circumstance of question 7 calculate the chemical energy of the explosive charge if the additional information is known that the cannon's temperature increases rapidly by 3°C. The specific internal energy per degree of the cannon material is 0·45 kJ/kg K.

[5 093 kJ]

Q. 9. Find the change in internal energy of water per kg when the temperature increases from 0 to 100°C assuming c_v = 4 kJ/kg K and water is incompressible. If this energy were supplied by heat to the system from a 3 kW immersion element for how long would the power have to be switched on?

[ΔU = 400 kJ, t = 133 s]

Q. 10. A rigid insulated vessel of negligible mass contains 10 kg of an explosive mixture at 25°C. If the energy released is 2 MJ/kg and the value of c_v = 0·75 kJ/kg K how much heat or work energy is released and what is the final temperature of the contents?

[Heat and work energies = 0. Final temperature = 2 692°C]

Q. 11. On a flat motorway compare (a) the energy dissipated in a head-on collision between two cars of mass 600 kg each, which are travelling in opposite directions at 100 km/h, if they are both stopped completely by the impact, with (b) the energy dissipated in the act of collision of two 600 kg cars of which the first was initially travelling at 60 km/h but hit the central crash barrier with a resulting loss of 10 per cent of its kinetic energy before returning onto the correct carriageway where the collision occurred between it and the second 600 kg car travelling at 70 km/h in the same direction. After the collision the two cars became locked together and moved forward at a joint speed of 60 km/h.

[463; 21·2 kJ]

4

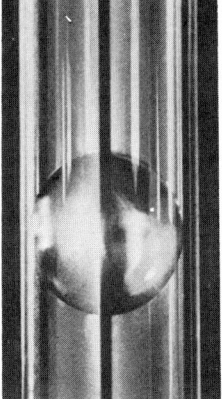

Energy conversion

Now we show how it is possible to interchange energy between the various forms of stored energy without there being any overall change in the total energy stored in the system. However, first we must discuss a point which causes some difficulty, the transfer of energy by work directly into the system.

4.1 Transfer of energy by work

Many students of thermodynamics find it difficult to understand the thermodynamicist's view of what happens at the boundary of a system when energy is transferred into the system from the surroundings by work. Consider a metal block rubbed without lubrication across a metal plate between positions C and D (see Fig. 4.1). Let the hand that does the rubbing be one system, the block a second system and the plate a third system.

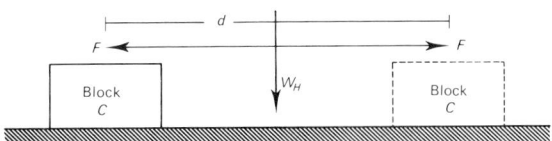

FIG 4.1   A block rubbed across a rough plane

The system comprising the hand does work W_H by moving against a force F through a distance d. The chemical and other changes in the hand associated with the work it is doing are complex but one thing is clear—the work it does causes a reduction of the hand's internal energy and the hand, if this work continues becomes tired.

The second system, the block, is best considered as nearly rigid and completely elastic. The energy W_H passes through the block and arrives at the boundary where it splits into two parts. One part, W_P, say, is transferred by work from the block to the plate. The rest of the energy, ΔU_B say, does not actually leave the block and therefore cannot be said to have been transferred out of the second system, the block. From the first law of thermodynamics energy must be conserved, so provided that are no heat losses to the surroundings, an **adiabatic** process

$$W_H = \Delta U_B + W_P \qquad\qquad (4.1)$$

The originator of the energy W_H was the hand and this energy has been transferred from the hand into the block by work and passes through the block until it reaches the block/plate interface. There it splits into two parts (see Fig. 4.2) of which one part, ΔU_B, remains in the block, and the other part, W_P, is transferred by work from the block to the plate.

The energy ΔU_B that remains in the block is observable by a rise in temperature of the block. At the same time the energy W_P that is transferred to the plate causes a rise ΔU_P of internal energy (see Fig. 4.3) in the plate (in fact $\Delta U_P = W_P$) observable by a rise in temperature of the plate. Heat has not been done in this elementary consideration of the problem. It is the energy put into the block by the hand that causes deformation of the crystals at the common surface and has been partly transformed to internal energy within the block and

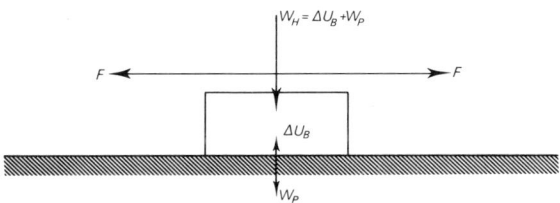

FIG 4.2   Division of work at the common surface

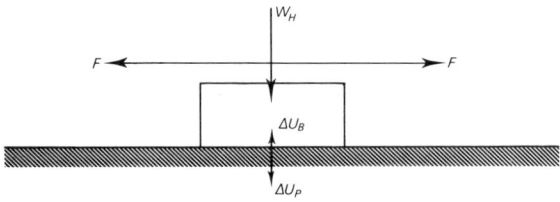

FIG 4.3   Increase of internal energy conducted from common surface

partly transferred by work out of the block into the plate where it is stored as internal energy. Some, in explaining the arrival of ΔU_B in the block, are tempted to say that the plate has done work on the block. This, from the thermodynamicist' point of view, is quite wrong as no energy has been transferred from the plate to the block so no work has been done by the plate on the block.

The point that causes surprise and often confusion in the minds of students of thermodynamics is that if the plate were thermally a non-conductor all the energy would remain in the block if this were relatively a good conductor. The equation (4.1) would become

$$W_H = \Delta U_B \text{ because } W_P = 0$$

If the converse were true and the plate was the relatively good conductor

$$W_H = \Delta U_P$$

If both were poor conductors the internal energy would be slow in getting away and the surface temperatures of the block and plate would rise and this could lead to the situation in which the two surfaces fused together.

The conclusion is a perfectly logical one, but will come as a surprise to most readers. When work is done on the common boundary of one system P and another B by, say, friction the work energy W_H is split into two parts ΔU_P and and ΔU_B which are conducted into the solids P and B in amounts directly proportional to the solids' thermal conductivities λ_P and λ_B. The rubbing causes a rise of temperature of the materials at the common surface. Although rubbing is a work action and is the primary action in this case, a secondary action—a heat action—will occur at the boundary which will tend to keep the common boundary temperatures of the solids the same. The dissipation of energy from the common surface into the bulk of the two materials will be given by Fourier's law of conduction

$$\Delta U_P = \lambda_P A_P \frac{dT}{dp}$$

and

$$\Delta U_B = -\lambda_B A_B \frac{dT}{db}$$

If $\dfrac{dT}{dp} = \dfrac{dT}{db}$ and $A_P = A_B$ because this is a common surface

$$\frac{\Delta U_P}{\Delta U_B} = \frac{\lambda_P}{\lambda_B} \qquad\qquad (4.2)$$

The thermodynamicist likes the above picture of what happens because it is logical and consistent and each word used has only one technical meaning. The confusion has been caused by statements such as 'heat is generated' instead of 'energy is transferred' or 'energy is transformed', and 'heat conduction' instead of 'energy conduction'.

After time has brought nearly steady conditions back to the systems the internal energies that have caused rises of temperature in both block and plate in the example discussed above are no different from having come into the systems by work or by heat. If they came by rubbing, they came by work; if they came from a hot source being placed next to them, they came by heat.

Now, having cleared ourselves, we hope, of any danger of the word 'heat' being used to represent energy let us return to the subject of energy conversion.

The block and plate (Q and A)

Q. 1   If conditions for the block and plate are as described above, and the ratios of the thermal conductivities of the material of the block to that of the material of the plate, λ_B/λ_P is $3/1$, what is the initial value of the ratio $\Delta U_B/\Delta U_P$?

A. 1 $\Delta U_B/\Delta U_P = 3/1$

Q. 2.   Your answer to question 1 refers to intial conditions. What happens later?
A. 2.   What happens later depends on the specific internal energy per degree, c_v, of the materials of the block and plate. These will decide the difference of surface temperature between the block and plate at their common surface. If there is a temperature difference then as a **secondary effect** heat action will occur at the boundary where hitherto there has been only work.

4.2   Conversion between K kinetic and Z gravitational energies

Consider a system comprising both a perfectly elastic ball and a tube with closed ends, within which the ball is bouncing, and assume for simplicity that the tube contains no air. The walls, the top, and the bottom of the tube are thermally and electrically non-conducting and they are rigid and opaque. The interior surface of the walls is quite smooth (see Fig. 4.4).

Because the walls are rigid the tube cannot transfer energy out by movement and because they are electrically non-conducting energy cannot be transferred out electrically. Therefore $W = 0$. Because the walls are thermally non-conducting and opaque (no radiation possible), energy may not be stored in the walls nor can it be conducted or radiated to the outer surface of the walls, therefore $Q = 0$. Energy transfers between the system and its surroundings are therefore impossible.

The ball falls from position A where it was stationary to position B (just before bouncing), and in doing so loses ΔZ_B of gravitational energy. But energy must be conserved and hence, in falling from A to B, the ball gains kinetic energy ΔK_B, and in terms of equation (3.2)

$$0 = \Delta K_B + \Delta Z_B \qquad (4.3)$$

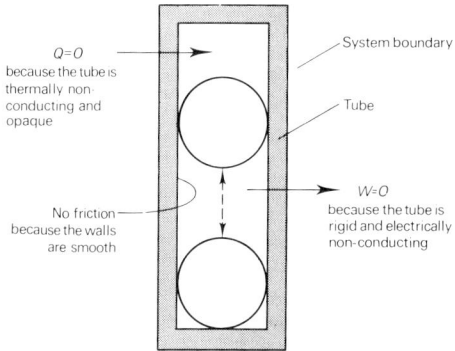

FIG 4.4   A ball bouncing within a closed smooth tube

The ball is also perfectly elastic. The consequence of this is that whenever the ball bounces it is deformed by pressure between it and the bottom closed end of the tube. Its elasticity returns it completely to its original shape when the pressure is relaxed. The sequence of events is that the ball moves from position B and strikes the bottom of the tube. The ball, thereby brought to rest, deforms and its energy, which was all kinetic energy before impact, is stored temporarily as elastic strain energy, as in the compression of a spring. Because of its perfect elasticity the ball returns precisely to its original shape and so converts back all the strain energy to kinetic energy without loss. The bouncing completed, the ball rises again to A, the kinetic energy being lost and the gravitational energy regained.

4.3   Conversion between U internal, K kinetic and Z gravitational energies

Now consider the same system of tube and bouncing ball, in which most features are still the same—the ball still being perfectly elastic—but with the difference that now the surface of the ball is rough. This causes friction between the ball and the tube if there is relative movement between them. As in the first case, because the tube's walls are thermally non-conducting and opaque no energy can be transferred into or through them by conduction or radiation (see Fig. 4.4). However the friction between the ball and the tube will cause the values of K and Z to be decreased as the ball bounces up and down. This is so because there will be another type of energy conversion at the places of contact between the ball and the tube. This conversion is within the system, as energy cannot be transferred anywhere by work because the tube is rigid and electrically a non-conductor, nor can it be conducted or radiated into the walls of the tube because

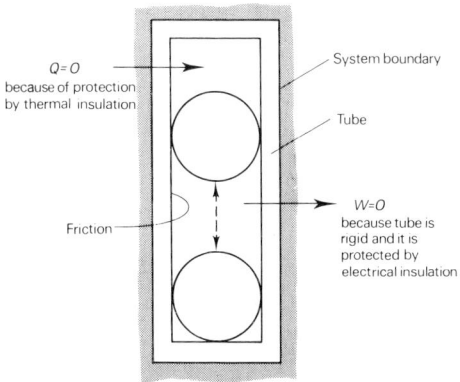

FIG 4.5   A ball bouncing within a closed rough tube

they are thermally non-conducting as well as opaque. If the ball is a good conductor then all the energy transferred out to the rubbing of the ball on the tube is conducted into the ball and increases the ball's internal energy, which would be observable by a rise of temperature of the ball. In the continuous exchanges of different
types of energy during the process the total energy must be conserved, and at any
time equation (3.2) must apply in the form

$$0 = \Delta U_T + \Delta U_B + \Delta K_B + \Delta Z_B \qquad (4.4)$$

In the case of the tube being a non-conductor no energy enters the tube walls so
$\Delta U_T = 0$, where U_T is the internal energy of the walls.

The bouncing ball (Q and A)

Q.1. If the ball has not been perfectly elastic what effect would this have had on
the energy changes?
A.1. When the ball bounces on the bottom of the tube some of the kinetic energy
will be used to deform the ball plastically and permanently. Because the ball
is permanently distorted it does not return precisely to its original shape and so
does not convert back all the strain energy to kinetic energy. There is an 'apparent
loss' of energy.
Q.2. Where does this 'apparent loss' of energy go?
A.2. It stays inside the ball as part of the ball's internal energy causing a rise
in temperature or if the impact is very severe it would appear as a permanent
deformation. If it appears only as a rise of temperature

$$\text{loss of kinetic energy} = Mc_V\Delta T$$

where M is the mass, c_V the specific internal energy per degree and ΔT the temperature rise of the ball.

Q.3. In section 4.3 we discussed conditions in which the ball was perfectly elastic but there was rubbing on the walls. In questions 1 and 2 we are discussing conditions in which the ball was not perfectly elastic but there was no rubbing on the walls. Give one similarity and one difference between these two cases.
A.3. One similarity is that the energy conversion is in both cases by deformation of the ball.
     A difference is that in the first case the deformation is of the crystal near the area of the surface affected by rubbing; in the second case the deformation is caused by a force deforming the ball.

4.4 The tube becomes a conductor

A more realistic situation would be that in which the walls, although still rigid, are thermal conductors and electrical conductors. The outside surfaces of the tube are however covered by some good insulating material (see Fig. 4.5).

As the ball falls, not only will the friction cause the temperature of the ball to rise as in the case of section 4.3 but also the temperature of the wall. In section 4.3 we wrote that, because the tube walls are thermally and electrically non-conducting and opaque, no energy can be transferred into or through them by conduction or radiation. The difficulty of non-conduction into the walls has now been removed and some of the energy converted by rubbing will go into increased internal energy of the tube instead of all going to increase the internal energy of the ball. If we assume that the same amount of enrgy has been converted by rubbing as before $\Delta U_T + \Delta U_B$ will in total have the same magnitude, but now ΔU_T will not be zero and ΔU_B will be smaller than before. The form of equation (4.4) will be unchanged.

4.5 Questions for the reader

Q.1 What changes of energy take place in a lump of soft putty dropping in a vacuum onto a rigid surface?

$$[\Delta Z \rightarrow \Delta U]$$

Q.2 Taking a banana as the system what changes of energy take place when a man squashes it with his foot?

$$[W \rightarrow \Delta U]$$

Q.3 A child places a nut on the surface of some treacle in a tin. It slowly sinks. Consider the nut before it has reached a steady velocity of sinking. What changes of energy are taking place so far as the nut alone is concerned?

$$[\Delta Z \rightarrow (\Delta U + \Delta K + W)]$$

Q.4. Consider the process of running a comb through one's hair so that the comb becomes electrostatically charged. What changes of energy are taking place if the system considered is

    (a)   The hand holding the comb
    (b)   The comb
    (c)   The hair?

    [(a), $\Delta U_{Hand} \rightarrow W$; (b), $W \rightarrow \Delta U_{Comb} + W_{Hair}$; (c), $W_{Hair} \rightarrow \Delta U_{Hair}$]

Q. 5.   On what equation in the text are all the above answers based?

    [Equation (3.2)]

Q. 6.   The system is an aeroplane, together with its passengers, that has landed in the desert because it has run out of fuel. What changes of energy are taking place so far as the system is concerned?

    [Q (from the sun) $\rightarrow \Delta U$]

Q. 7.   A pendulum swings back and forward immersed in a swimming bath. If (a) the pendulum is the system and (b) the water is the system what energy changes occur?

    [(a)   Intermittent $\Delta Z \leftrightarrow \Delta K$, $(\Delta Z + \Delta K) \rightarrow (W + \Delta U_{Pend})$;

    (b) $W \rightarrow \Delta U_{Water}$]

Q. 8.   Water at 20°C flowing at 10 m/s, falls down a waterfall 15 m high. If c_V of the water is 4.2 kJ/kg K calulcate the temperature of the water at the bottom of the fall if it flows away with a velocity of 10 m/s.

    [20·035°C]

Q. 9   If the water at the bottom of the fall in question 8 flows away with negligible velocity what is its temperature?

    [20·047°C]

Q. 10   If a 4 kg cake were dropped from a cliff 100 m high and all the energy was converted to kinetic energy what would (a) be the velocity of the cake before impact, (b) where does this energy go to!

    [44 m/s, the energy goes into breaking up
    the cake as it hits the rocks!]

5 Availability

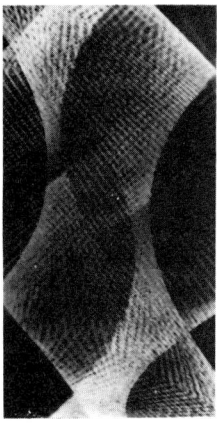

This chapter considers the level of energy and explains that, although the sum of all the energy in the Universe is assumed to remain the same (in this classical sense), it is nevertheless being continually degraded. This idea of the level of energy, its availability, is used to introduce the second law of thermodynamics.

5.1 Conservation of energy

Heat and work may occur at the boundary of a system undergoing a process. Equation (3.2) is a statement of the balance of energy in a system undergoing such a process.

$$Q - W = \Delta U + \Delta K + \Delta Z \tag{3.2}$$

It states that the sum of the energies Q entering a system by heat and W leaving the system by work during a process is equal to the sum of the changes of internal energy U, of the kinetic energy K, and of the gravitational energy Z. We now wish to discuss the conservation of energy undergoing the two types of transfer occurring at the system boundary: (1) The action of heat.   (2) The action of work.

Consider an energy transfer Q by heat from a system which is a block of metal A (see Fig. 5.1 (a)) at temperature T_1, in thermal contact with its surroundings, represented in the figure by another metal block B at temperature T_3. Because of the difference in temperature there will be an exchange of energy between A and B. If T_1 is a higher temperature than T_3 energy will be transferred by heat from A to B. Because we are dealing with stationary metal blocks the changes ΔK and ΔZ in the kinetic and gravitational energies will be zero and only changes ΔU in the internal energy of the blocks will occur. The energy will continue to be transferred in the direction of the arrow in Fig. 5.1 (a) until the temperatures of both metal blocks are the same at, say, T_2 (see Fig. 5.1 (b)). T_1 will

38

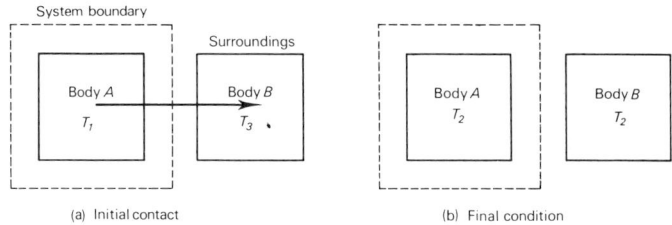

(a) Initial contact   (b) Final condition

FIG 5.1   Two objects at different temperatures in thermal contact

have fallen to T_2, and T_3 will have risen to T_2—during this process there will have been an energy transfer Q by heat from the system block A to the surroundings represented by block B.

To evaluate the changes of internal energy of each block, they must each be considered as a system and then we make use of a form of equation (3.3):

$$\Delta U = c_V M \Delta T \qquad (5.1)$$

where U, M, c_V and T are the internal energy, mass, specific internal energy per degree and the temperatures of the block systems. Now let us use the first law of thermodynamics, applying it to the combination of A plus B knowing that during the process there is no energy transfer from this combination of system and surroundings. The the internal energy initially in A plus that initially in B are together equal to that finally in A and B put together. That is to say:

$$c_A M_A (T_1 - T_0) + c_B M_B (T_3 - T_0) = c_A M_A (T_2 - T_0) + c_B M_B (T_2 - T_0)$$
$$(5.2)$$

We are thereby stating the principle of the first law—that energy must be conserved. The final temperature T_2 of the block can be found if the initial conditions and material properties are known. In equation (5.2) the temperature T_0 is a datum for temperature.

Energy transferred by work is also conserved. Consider work energy W, and let the system be a block of metal M (see Fig. 5.2) in mechanical contact by means of a rope with its surroundings, represented in Fig. 5.2 by a continuation of the rope, a windlass, an electric motor and a power supply. Because the motor turns the windlass there will be an exchange of energy between the system and its surroundings. If the motor is of 1 kW capacity and operates for 10 s in raising the block against gravity energy will be transferred from the power supply to the block. If we are dealing with a block being raised slowly the changes ΔK and ΔU in the kinetic and internal energies will be zero and only changes ΔZ in the gravitational energy of the block and the expenditure of electric energy in the surroundings need be considered. The energy will continue to be transferred in the direction of the arrow W in Fig. 5.2 until the height of the block is increased by ΔZ, by which time energy W will have been transferred.

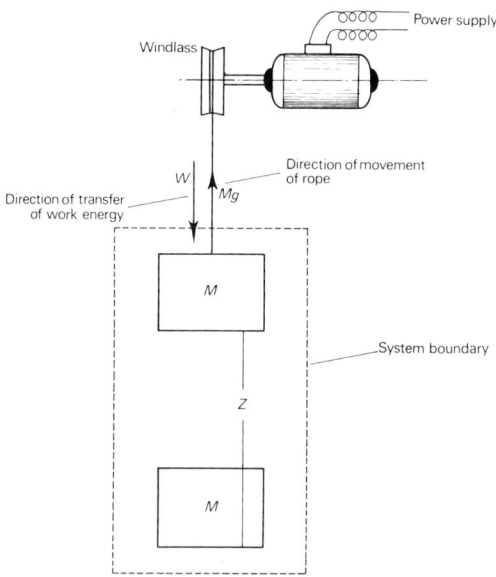

FIG 5.2   The raising of a mass

To evaluate the transfer of energy from the power-supply system, assuming losses are negligible, we multiply the power of the motor by the time and find that

$$W = \text{power} \times \text{time} = 1 \times 10 = 10 \text{ kJ} \qquad (5.3)$$

and to evaluate the change of energy in the system we make use of equation (3.5):

$$\Delta Z = Mg\Delta \bar{z} \qquad (5.4)$$

where Z, M, g and $\Delta\bar{z}$ are the gravitational energy, mass, strength of the gravitational field and height through which the block M is raised. We must again use the first law to state that energy is conserved, or in other words that the gravitational energy gain by M is equal to the energy transferred by work, or from equations (5.3) and (5.4),

$$-(-W) = W$$
$$= 10 \text{ kJ from equation (5.3)}$$
$$= \Delta Z \text{ from equation (3.2)}$$
$$= Mg\Delta\bar{z} \text{ from equation (3.5)}$$

5.2   Loss of availability during heat transfer

In the heat transfer shown in Fig. 5.1 all the energy is conserved. By this we mean that there is the same total energy in blocks A and B in the final situation, shown in Fig. 5.1 (b), as there was in blocks A and B in the initial situation shown in Fig. 5.1 (a). There has, however, during the course of the process, been a redistribution of the energy between blocks A and B. The energy initially in store in block A was at temperature T_1 and that initially in store in block B was at temperature T_3. Then, during the process, there was an energy transfer by heat so that in the final situation all the energy in blocks A and B is at temperature T_2. We say there has been a loss of availability.

To understand what is meant by loss of availability one must examine the usefulness of the energy stored within a system. In this case the energy in block A was the most useful, the most available of all we are saying. Some of it could be of service in the course of being transferred by heat to another store at any temperature lower than T_1. Now look at the final situation. In the final situation the energy in A can be of service during a heat action only to stores that have a temperature lower than T_2. In the first situation some of the energy could be transferred by heat to any system whose temperature is lower than T_1 but, after the process has occurred, none of the energy can be transferred to any system whose temperature is higher than T_2. This is the extent to which the energy becomes less available during this process involving energy transfer by heat between blocks A and B. If the system, block A, and the surroundings, block B, are considered, then something that could be done before the process took place can no longer be done after. It all follows because energy cannot be transferred by heat alone from a colder to a hotter body. It is true of course, if block B is considered as the system, that during the course of the process its temperature has risen from T_3 to T_2. However one must consider both the system and the surroundings before saying whether, overall, there is a gain or loss of availability. So this leads us to conclude from this process of heat transfer that initially some energy could be transferred from a store with a maximum temperature of T_1 and finally from a store with a maximum temperature of T_2. Some energy therefore has been degraded permanently.

5.3   The first and second laws

What was stated in section 5.2 about availability is the principle behind the second law of thermodynamics which states that

'When energy is exchanged by heat between a system and its surroundings some energy, overall, must be permanently degraded to a lower level, unless a reversible process occurs.'

The first law states that energy must be conserved and this we may therefore think of as the **LAW OF CONSERVATION.** The second law states that energy is continually being degraded and we can therefore think of this as being the **LAW OF DEGRADATION.**

5.4 Loss of availability during work transfer

Energy is stored in a system in the form of internal, kinetic or gravitational energy whether the action at the system boundary is work or heat. An example of energy being transferred into a system by work and stored as internal energy can be seen in the rise of temperature of a gun-barrel being bored by a boring tool. The action of removing metal from the bore of the gun-barrel by work has resulted in a rise in temperature of the gun-barrel because of an increase of its internal energy. This could have been produced by applying a flame to the gun-barrel but such a transfer would have been due to a difference of temperature between the barrel and its surroundings and would therefore have been heat not work.

What was stated in section 5.2 about loss of availability of energy transferred by heat, and the words used in the second law itself, appear at first sight to imply that loss of availability occurs only during heat and not work transfer.

This is only partly true. What is true is that there is no loss of availaibility of that part of the energy transferred by work and actually stored as gravitational energy as shown in Fig. 5.2. This is because such energy can be recovered as work energy and can be transferred in both directions between any two systems no matter what their temperatures may be. These considerations also apply to energy transferred by work and stored as kinetic energy. There is however loss of availability of energy transferred by work and stored after transfer as internal energy at a particular temperature. It can then only be transferred by heat to a lower temperature

Availability (Q and A)

Q.1. A man has four buckets, each containing 5.5 kg of water. The water has a constant c_v of c kJ/kg K and a temperature of 50°C in two buckets and of 10°C in the other two. The man also has a bath tub of adequate capacity in which he intends to use no less than 15 kg of water at 38°C, these being the maximum quantity and minimum temperature that he finds satisfactory for a bath. Is there sufficient energy available to him?

A.1. He requires 15 kg at 38°C. Suppose that he uses x kg of cold water and y kg of hot water in producing his requirements. The energy balance would, from equation (5.2), be:

$$y \times c \times (50 - 0) + x \times c \times (10 - 0) = (x + y)c \times (38 - 0)$$

when a datum of zero energy is assumed at T_0 where $T_0 = 0$°C, the mass balance is given by $x + y = 15$. From these two equations $x = 4\cdot5$ and $y = 10\cdot5$ kg. There is sufficient water available to give the quantity required at the right temperature, so sufficient energy is available.

Q.2 In a fit of absent mindedness the man preparing his bath empties the contents of all the buckets into the tub. Having in mind the first law of thermodynamics, the conservation law, has he (a) enough stored energy now in the water to make a satisfactory bath, and (b) enough available energy for the purpose?

A.2   (a) Yes, the first calculation showed there was sufficient energy to prepare the bath and the first law states that this is conserved.
(b) No, all the water is now at 30°C and he cannot now prepare his bath at 38°C—in fact the energy, being all at a lower temperature than before is less available to him than it was.

5.5   Summary

The availability of energy has been discussed and changes in availability of energy were examined when systems undergo processes in which the actions of heat and work occur; this led to a statement of the second law of thermodynamics.

5.6   Questions for the reader

Q.1   A boy scout rubs two sticks together. If the system is the two sticks, has the availability of the energy transferred between the system (the two sticks) and its surroundings been increased or decreased during the process?

[Decreased, because energy was available at any level before it was transferred by work to storage as internal energy]

Q.2   A stirrer is used to whip a milk shake which is maintained at constant temperature by immersing the milk-shake container in a mixture of ice and water. If the system is the milk shake only, how has the availability of the energy transferred during the process changed?

[Decreased, similar to question 1. In this case the surroundings, the ice and water, have increased internal energy]

Q.3   An inventor claims that the possibility is worth investigating of using an artesian well of large capacity as the only source of energy to drive a generator for a district lighting scheme. Is he right?

[Yes, the artesian well would do work on the lighting system and therefore the second law would not be broken]

Q.4   A large mass of metal is dropped down a well shaft and in so doing is used to drive a turbine. Is this possible and would you advise it?

[It is possible because it is proposed that the energy input and output are both by work, and so no law is broken—but the cost of the project might by disproportionate to the advantage gained]

Q.5.   A space-craft descends to Earth through the atmosphere. The hot junctions of a thermopile (many thermocouples) are fitted to the heat shield, and cold junctions to the inside of the air-conditioned capsule. The electrical energy from the thermopile system is used to operate a guidance control mechanism. Is this an intelligent thing to try?

> [Yes, work done between the capsule and the atmosphere will produce a temperature difference between the ends of the thermopile producing electrical energy which can be used to drive the guidance mechanism

Q. 6.   A 50 kg mass of steel of specific internal energy per degree of 0.42 kJ/kg K at 500°C is dropped into 5000 kg of liquid at 38.6°C with a specific internal energy per degree of 2.0 kJ/kg K. For what range of temperature is the energy still available?

> [Below 39.6°C]

Q. 7.   In a fish-farming experiment the cooling water from a nuclear-power station is used. Water is taken from the sea at 10°C and is returned to a fish tank at 30°C after passing through a condensor used in the nuclear plant. Each 5000 kg of this warmed water, which is one hour's use, is mixed with some additional cold sea-water so that the temperature of the mixture is 20°C. The specific enthalpy per degree of water is 4.2 kJ/kg K. Calculate

> (a)   the flow-rate of the additional cold sea-water to produce the 20°C mixture.
> (b)   the energy that becomes unavailable for transfer to a thermal resevoir at (i) 10°C and (ii) 25°C.

> [a) 5000 kg/h; b) (i) None, (ii) 105 000 kJ/h]

Q. 8   Steam is the working fluid in a system that comprises boiler, turbine and feed pump. Energy is transferred from outside the system through the walls of the boiler. The steam then does work in the adiabatic turbine (the word adiabatic means no transfer by heat, i.e. Q for the turbine = 0) and afterwards fed by the feed pump back to the boiler. Is the availability of the energy transferred decreased or increased? Is the system capable of working as a steam turbine plant and why?

> [Increased, not capable, because unless the feed pump does more work than the turbine—an absurd state of affairs for a steam turbine plant—the system would break the second law in that it would be doing positive work while exchanging heat energy with a single reservoir]

6  *Heat engines*

The first and second laws of thermodynamics are applied to a heat engine which is the traditional name given to a system that converts heat energy into work energy. Various ratios are considered that measure the performance of the heat engine.

6.1 A heat engine

The magic box mentioned in Chapter 1 as being sought by Man to do his work would most likely be supplied with energy by heat, because nature provides plentiful resources for this. The box would in response supply energy by work. From what was said in Chapter 2 about a system it should be clear that such a magic box is a system, and indeed it is a system undergoing a special sort of process. It is for this reason that we give this system the special name of heat engine (see Fig. 6.1). It will be seen in the figure that heat energy Q_1 is supplied to the engine and work energy W is supplied by it.

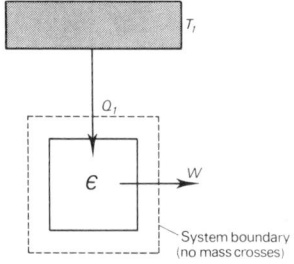

FIG 6.1 A proposed heat engine

But if we apply the second law to the engine in Fig. 6.1 we find something wrong in that the heat energy Q_1, available at temperature T_1 and being transferred to the system by heat, is being transferred out of the system, undiminished,
by work. Because it is all being transferred out by work it could consequently all
be stored by the raising of a mass—see definition of work—and could therefore all
be recovered at any level of availability (see section 5.4). This, according to the
second law, is impossible because during a process involving a heat transfer there
must be a permanent degrading of some of the energy (see section 5.3). For a
cyclic process (see Chapter 2) additional energy transferred to the system can
only be stored for periods of less than one cycle. It follows that some of Q_1—let
us say an amount Q_2—must be transferred out of the system and stored elsewhere
at a level of availability lower than T_1. The conclusion reached is that, for a heat
engine to be possible we must make provision for the incoming heat energy Q_1
to be divided into two outgoing parts,

(a) W which will be the work that we require, and
(b) Q_2 which will go to a lower level of availability.

It follows therefore that for an engine to be possible it must be like
that shown in Fig. 6.2 and, to clarify this point, Max Planck in 1897 wrote a version
of the second law specifically applying to the heat engine. He wrote

'It is impossible to construct a heat engine that will do positive work
while exchanging heat energy with a single reservoir'.

Expressed in this way, a way that follows naturally from the version
of the second law that we gave in Section 5.3 applied to the engines proposed in
Figs. 6.1 and 6.2, we find that the second law admits the possibility of the engine
shown in Fig. 6.2 but denies the possibility of the engine shown in Fig. 6.1. One

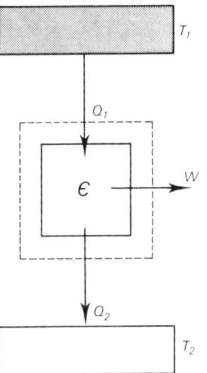

FIG 6.2   A possible heat engine

can argue that, if the low-temperature reservoir were at absolute zero—that is if it were in a condition of storing no energy—then no energy could be transferred to it and that when one is dealing with an engine using a low-temperature reservoir at absolute zero Q_2 can be zero. Absolute zero is a very complex conception and is not likely to come into the field of practical design. So, except for the special case of T_2 in Fig. 6.2 being absolute zero we have every reason to believe that Max Planck's version of the second law is true.

Later, in section 6.4, performance is defined as the ratio:

$$\frac{\text{What one gets that is useful}}{\text{What one pays to get it}}$$

Using this ratio, the performance of an impossible heat engine such as that in Fig. 6.1 is given by:

$$\text{Performance} = \frac{W}{Q_1} \tag{6.1}$$

The first law applied to the same engine gives

$$W = Q_1 \tag{6.2}$$

and (6.1) and (6.2) together give the performance of the engine as $W/Q_1 = 1$. The engine of Fig. 6.1, if it were possible, would therefore have an efficiency of 1 (i.e. 100 per cent).

6.2   A heat engine and its surroundings

Figure 6.2 shows two important items that are required as parts of the surroundings of the system called a heat engine. These two items are

(a)   a high-temperature reservoir T_1
(b)   a low-temperature reservoir T_2

The heat engine takes in energy Q_1 from the high-temperature reservoir by heat and sends it out again in two parts, W by work and Q_2 by heat to the low-temperature reservoir, and in accordance with the first law (see equation (2.1) the following energy balance is true,

$$Q_1 + Q_2 - W = 0 \tag{6.3}$$

When substituting numerical values in equation (6.3), note that Q_1 and W are positive (see section 2.2) and Q_2 is negative.

In this chapter we look into the engine and see the processes used to change Q_1 to W and Q_2. In Fig. 6.3 are shown all the basic processes required for this change. All the processes are undergone by a **working fluid**—in a steam power plant the working fluid would be water—that is pumped through a piped system

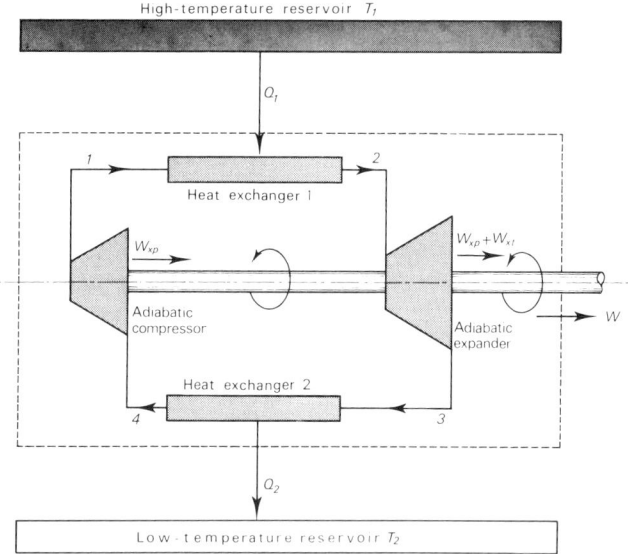

FIG 6.3   The processes within a heat engine

around the cycle 1 2 3 4 1 (see Fig. 6.3). Consider the working fluid to be in state 1 at point 1 of the cycle at which its temperature is less than the temperature T_1 of the high-temperature reservoir. From point 1 the working fluid is forced by pressure into the heat exchanger **1**[*] which is in thermal contact with the high-temperature reservoir. Because the temperature at 1 is lower than T_1 there will be an energy transfer Q_1 by heat from the high-temperature reservoir to the working fluid. In a steam turbine plant heat exchanger **1** would be a boiler or steam generator, and in a gas turbine plant it would be a combustion chamber. During the process the working fluid moves to point 2 from which it undergoes a different process in passing through an adiabatic expander—which might be a steam or gas turbine—doing useful work W_{xt} and consequently losing energy before arriving at point 3. We introduce the subscript x to indicate the work energy is coming from a control volume (see Chapter 8), not from a system. The loss of energy between points 2 and 3 may be observed as a drop in temperature and pressure of the working fluid. The working fluid then passes from 3 to heat exchanger **2** where the surroundings—the low-temperature reservoir—are at temperature T_2 (T_2 being lower than the temperature of the working fluid at 3). Between points 3 and 4 energy Q_2 will be transferred out of the system by heat, so that the working fluid leaves the

---

[*] A heat exchanger is the name given to a piece of equipment within which energy is exchanged by heat between a system and its surroundings.

heat exchanger **2** at its coldest—as cold as T_2 or nearly so. From points 4 to 1 the working fluid passes through an adiabatic compressor in which work is done on it, transferring W_{xp} to it and restoring its pressure and temperature to its original state at point 1. This description is of a series of four processes undergone by a working fluid within a system which, when considered as a whole, is seen to put the fluid through a composite cyclic process.

These four processes when considered together as a cyclic process have the following features,

(a)   the total mass of the working fluid remains constant,
(b)   on completion of the cycle the working fluid has returned to its original state.

Two processes involve energy transfers by work; these are the energy W_{xt} from the turbine to the surroundings, and W_{xp} from the surroundings to the compressor. The total useful energy transferred by work W is given by the following equation

$$W = W_{xt} + W_{xp} \qquad (6.4)$$

When numerical values of W_{xt} and W_{xp} are used, W_{xt} will be found to be positive and W_{xp} will be negative, and for a heat engine $W_{xt} > W_{xp}$ so that W is positive.

6.3   A reversed engine

When the heat engine of Fig. 6.2 is operated in reverse as shown in Fig. 6.4 it is called a heat pump or refrigerator, depending on its function. In this case W' units of work energy and Q_2' units of heat energy are taken into the system

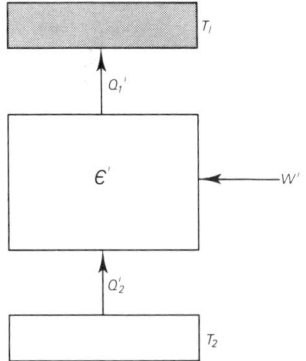

FIG 6.4   A reversed heat engine

from the surroundings while Q_1' units of heat energy are given out by the system to the surroundings. So from equation (2.1) we get the following energy balance:

$$Q_1' + Q_2' - W' = 0 \qquad (6.$$

When numerical values of Q_1', Q_2' and W' are used, Q_1' and W' are negative and Q_2' is positive. This may be compared with the case of the heat engine, see equation (6.3).

In Fig. 6.5 are shown the basic processes used in a reversed heat engine, and it can be seen by comparing Figs. 6.3 and 6.5 that the processes in a reversed engine are very similar to those of the forward-working engine of Fig. 6.3. Again the working fluid is pumped through a piped system in the same direction around the cycle 1 2 3 4 1. If we suppose the working fluid in state 1 is at a higher temperature than the temperature T_1 of the higher temperature reservoir we understand why the energy Q_1' leaves the working fluid within heat exchanger 1 and is transferred to the high-temperature reservoir. Heat exchanger 1 in a reversed engine might be a condenser. The working fluid in moving from state 2 to 3 passes through an adiabatic expander which in a reversed engine is often a throttle in which the working fluid undergoes an expansion to the lower pressure of point 3. The working fluid is then in state 3 at point 3 where its temperature is lower than the temperature T_2 of the low-temperature reservoir.

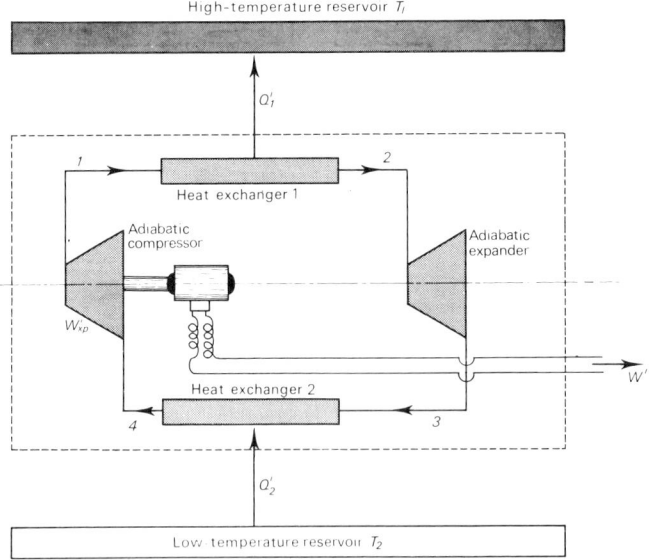

FIG 6.5   The processes within a reversed heat engine

In passing from point 3 to 4 the working fluid passes through heat exchanger **2** in which energy Q_2' is transferred into the working fluid by heat.

The fluid leaves the heat exchanger **2**—which in a refrigeration plant is often an evaporator— at point 4 in which state its temperature is a little lower than T_2. The fluid then passes through an adiabatic compressor in which work energy W_{xp} is transferred to the working fluid in restoring it to its original state at point 1.

These four processes again combine to produce a cyclic process in which

(a) the total mass of the working fluid is constant,

(b) on completion of the cycle the working fluid has returned to its original state.

The energy transfer by work during the whole cycle is, in this case, only W_{xp}' and therefore

$$W' = W_{xp}' \qquad (6.6)$$

where W_{xp}' is negative and hence W' is also negative.

Processes (Q and A)

Q. Draw the diagram corresponding to Fig. 6.3 for a steam plant consisting of a boiler, a turbo-alternator, a water condenser and a feed pump. Assume the feed pump and the turbo-alternator lie on the same shaft. For the answer see Fig. 6.6

6.4 Performance

In order to compare different heat engines some type of assessment of each engine is required, and for this purpose we define **Performance** as the ratio

$$\frac{\text{What one gets that is useful}}{\text{What one pays to get it}} \qquad (6.7)$$

In the case of the forward-working engine we call the ratio **Efficiency** and in a reversed engine we call it **Coefficient of performance.** As will be explained later there are two coefficients of performance, depending on whether one is using the reversed engine as a refrigerator or as a heat pump.

Let us take the simple example, shown in Fig. 6.3, of the forward-working engine. The useful energy we get from the engine is the work energy W and what we have to pay for it is the heat energy Q_1—we assume for this purpose

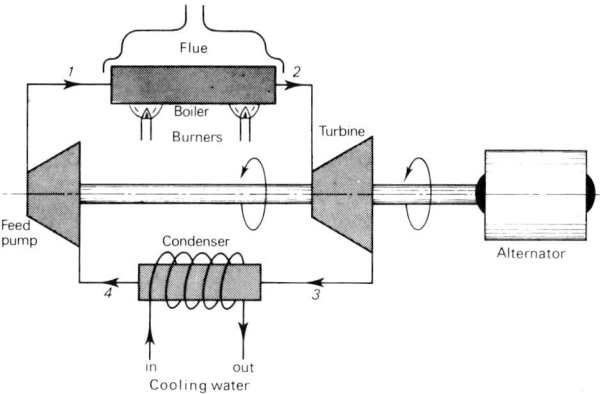

FIG 6.6   Answer to question about processes

that Q_2 is totally wasted. Therefore, from equation (6.7), we can say that the efficiency of the engine of Fig. 6.3 is given by

$$\eta = \frac{W}{Q_1} \tag{6.8}$$

$$= \frac{Q_1 + Q_2}{Q_1} \quad \text{from equation (6.3)}$$

$$= 1 + \frac{Q_2}{Q_1}$$

When substituting in values for Q_1 and Q_2, Q_1 will be positive, because it is heat energy entering the system, and Q_2 will be negative, because it is heat energy leaving the system.

A designer has many things to consider with regard to an engine such as its mass, the accessibility of its components, as well as the engine's efficiency. When considering its efficiency his aim is to make, for any value of Q_1, the efficiency η as large, and therefore Q_2 as small, as possible.

Efficiency (Q and A)

Q. If Q_1 of the engine in Fig. 6.3 were 1 800 kJ and Q_2 were −800 kJ what would be the engine's efficiency?

A. From equation (6.3)

$$W = Q_1 + Q_2$$

$$= 1800 - 800$$

$$= 1\,000 \text{ kJ}$$

From equation (6. 8)

$$\eta = \frac{W}{Q_1} = \frac{1000}{1800} = 0\cdot56$$

or 56 per cent

Another simple example, shown in Fig. 6. 5, is that of the reversed engine. The useful energy we get from the engine when it is used as a heat pump is the heat energy Q_1', and what we have to pay for it is work energy W'. Therefore, from equation (6. 7), we can say that the coefficient of performance of the heat pump of Fig. 6. 5 is given by

$$C_{hp} = \frac{Q_1'}{W'} \tag{6.9}$$

$$= \frac{Q_1'}{Q_1' + Q_2'}$$

6.5 Summary

Arising from Man's wish to have a dependable moveable device, that would feed on what is plentiful and in return do his work for him, was developed the heat engine—this takes in heat energy and gives out work energy. Nevertheless the heat engine is subject to the second law of thermodynamics and, while the engine is working, energy is being continually degraded. As a consequence it is seen that the engine cannot do useful work energy and at the same time exchange heat energy with a single source. A reversed heat engine is also described and standards of performance are established for both the heat engine and the reversed engine.

6.6 Questions for the reader

Q. 1 Draw a diagram corresponding to that shown in Fig. 6. 5 for a domestic refrigerator consisting of a cooling element inside the unit, an electric pump, an element at the rear of the refrigerator, which loses energy and a throttle.
[For answer see Fig. 6. 7]
Q. 2. Draw a diagram corresponding to that shown in Fig. 6. 5 for the heating system of a building which consists of a series of radiators in the building, a throttle, an element in a nearby river gaining energy from the cold river water and a pump.
[For answer see Fig. 6. 8]
Q. 3. A ship's steam power plant consists of an oil-fired steam boiler and exhaust flue, an adiabatic turbine driving a propeller, a condenser cooled by desalinated sea-water and a feed pump. Draw a diagram for this plant corresponding to that in Fig. 6. 4.
[For answer see Fig. 6. 9]

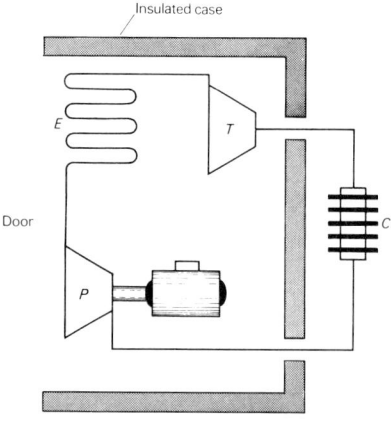

E = Evaporator or cooling element
P = Electric pump
C = Condenser
T = Throttle

FIG 6.7   Answer to Question 1

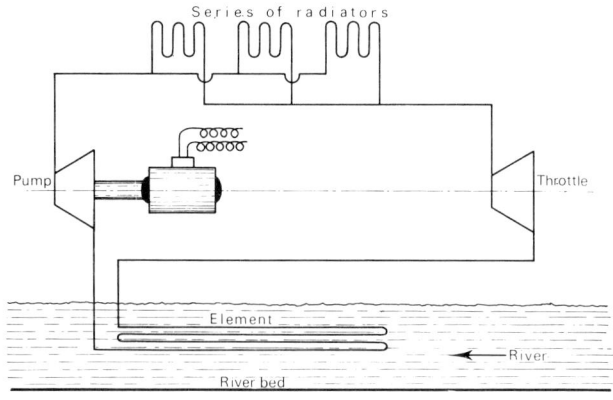

FIG 6.8   Answer to Question 2

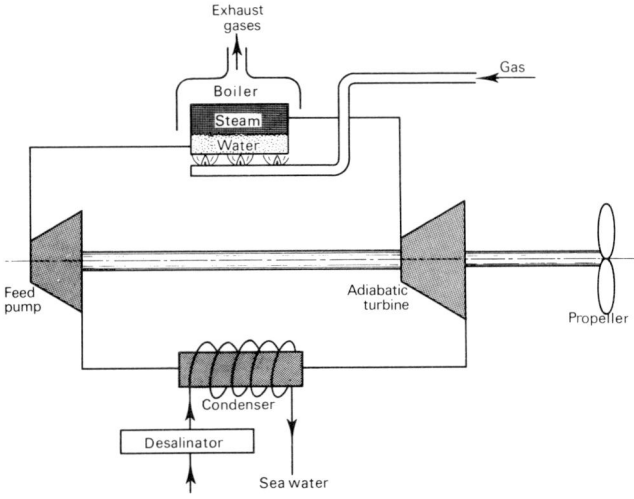

FIG 6.9   Answer to Question 3

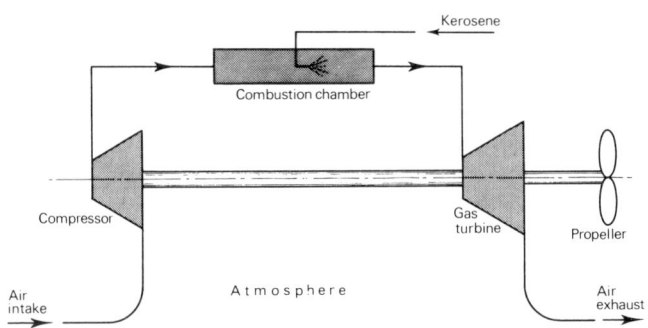

FIG 6.10   Answer to Question 4

Q. 4. Draw a similar diagram for an aeroplane's gas-turbine plant consisting of an air intake from the atmosphere, a compressor, a combustion chamber where kerosene is injected into the air stream, a gas-turbine driving a propeller, and an air discharge into the atmosphere.
[For answer see Fig. 6. 10]

Q. 5. A heat engine has a maximum work energy output of 3 kW. If the engine's efficiency is $0 \cdot 35$, calculate the heat energies Q_1 and Q_2. If, at half its maximum output, Q_1 falls to 5 kW what is its efficiency under these conditions?

$$[Q_1 = 8 \cdot 6 \text{ kW}, Q_2 = 5 \cdot 6; \eta_{5\text{kW}} = 0 \cdot 30]$$

Q. 6. The output of an engine is 15 kW while energy is being rejected through the radiator and other sources at the rate of 90 MJ/h. At what efficiency is the engine working?

$$[\eta = 0 \cdot 38]$$

Q. 7. The boiler of a steam-turbine plant takes in water at 100°C, and delivers steam to the turbine at 100°C. The energy of evaporation at 100°C from water to steam is 2 250 kJ for each kg of water evaporated. If the rate of evaporation is 2 kg/s at what rate in kJ/s does the boiler operate?

$$[4\,500 \text{ kJ/s}]$$

Q. 8. An engine that has a heat energy input of 75 kW has a heat energy output of $62 \cdot 5$ kW. What is its rate of working and what is its efficiency?

$$[W = 12 \cdot 5 \text{ kW}; \eta = 0 \cdot 17]$$

7   *Ideal heat engines*

      Engines with 100 per cent efficiency considered in the last chapter were seen to violate the second law of thermodynamics. So we must try to deal with the question, 'What, in terms of efficiency, is the best engine operating between two specific thermal reservoirs?'

7.1   A 100 per cent efficient engine

      For an engine such as that shown in Fig. 6.2 to be 100 per cent efficient, the work energy W must be the same as the heat energy Q_1 or,

$$W = Q_1 \qquad\qquad (7.1)$$

The first law applied to the engine of Fig. 6.2 states that

$$Q_1 + Q_2 - W = 0 \qquad\qquad (7.2)$$

It follows from equations (7.1) and (7.2) that, in a 100 per cent efficient engine, $Q_2 = 0$ and therefore such an engine would be like those shown in Figs 6.1 and

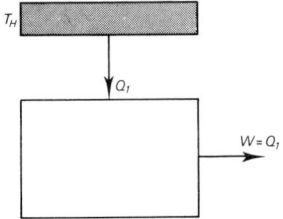

FIG 7.1   A 100 per cent efficient engine

7.1 and unlike that shown in Fig. 6.2. According to Planck (see section 6.1) Q_2 cannot be zero if W is positive, and if we believe Planck we believe the engine in Fig. 7.1 is not possible except perhaps in the impractical case in which T_2, the temperature of the low-temperature reservoir, is absolute zero—more will be written about this in a later chapter. Nevertheless the principles behind the possibility of having a 100 per cent efficient engine are worth further study.

7.2 A 100 per cent efficient engine reversed

If the 100 per cent efficient engine, such as that shown in Fig. 7.1 were reversed it would be like that shown in Fig. 7.2 taking in work energy W' and delivering heat energy Q_1'. As it is defined as having a 100 per cent performance we know from the first law that $W' = Q_1'$.

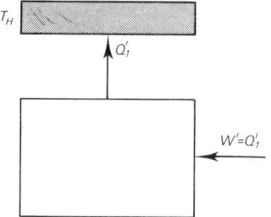

FIG 7.2   A 100 per cent efficient reversed

The energy W' being taken into the reversed engine is work energy and could, by definition, be stored wholly by the raising of a mass and be transferred into a reservoir at any temperature. However, in this case, it is being transferred into the engine by work and is subsequently transferred in the form of heat energy Q_1' to the reservoir at temperature T_H—it could indeed have been transferred by heat to a reservoir at any temperature if the right system and surroundings were considered.

Before the energy was transferred into the reversed engine of Fig. 7.2 it was available for transfer by a suitable reversed engine to a reservoir at any level because it was then stored in the form of gravitational energy, but after going through our reversed engine it is stored in a reservoir at temperature T_H. It is now available for transfer by heat to other reservoirs at temperatures lower than T_H. This is consistent with our statement of the second law which state that if energy is transferred by heat, as Q_1' in Fig. 7.2 undoubtedly is, then some energy must be permanently degraded. Before it was transferred by work into the reversed engine it was available for transfer to reservoirs at any temperature. Now it is available for heat transfer only to reservoirs at temperatures lower than T_H.

The engine of Fig. 7.2 is a reversed engine that we know to have a performance of 100 per cent, because $Q_1' = W'$. Nothing said in the second law precludes a 100 per cent performance for a reversed engine. Also, Planck's restriction of the performance of an engine refers only to an engine doing positive work. It does not refer to this case in which the work is negative. Because energy stored, or able to be stored, in the form of gravitational energy can be transferred to a reservoir at any temperature there are no restrictions on its availability.

7.3   Confirmation of Planck's version

Planck's version of the second law is another way of stating the impossibility of a 100 per cent forward-working engine in a situation governed by the version of the second law given in section 5.3. To show that a 100 per cent forward-working engine is an impossibility if the section 5.3 version of the second law is true let us suppose:

(a)   that the section 5.3 version of the second law is true, and
(b)   that a 100 per cent forward-working engine is possible.

Are these two possibilities compatible? Consider the 100 per cent forward-working engine shown in Fig. 7.3 taking in energy Q from a reservoir at temperature T_L

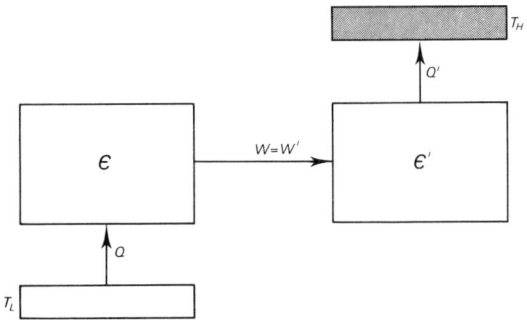

FIG 7.3   An engine and a reversed engine coupled together

and delivering it without loss in the form of work energy W. Applying the first law to this engine

$$W = Q \tag{7.3}$$

The work energy W becomes the energy intake W′ to a 100 per cent engine reversed. From section 7.2 we know that a 100 per cent reversed engine is possible. It takes in work energy W′ and delivers heat energy Q′. Applying the first law to this engine we get

$$W' = Q' \tag{7.4}$$

the numerical values of W' and Q' both being negative. From equations (7.3) and (7.4) we find that if we make the magnitudes of W and W' the same then the magnitudes of Q and Q' are the same. In this way the energy Q is seen to have been transferred from a low to a high temperature reservoir without any net negative work energy. This contravenes the second law as stated in section 5.3. We have already accepted the possibility of a 100 per cent reversed engine and also the second law statement of section 5.3. Therefore it must be that part of our argument that is at fault. This could be nothing other than our acceptance of a 100 per cent forward-working engine. Such an engine is not consistent with the section 5.3 statement of the second law, or with Planck's statement about forward-working engines. So we will conclude that the only possible forward-working engine is that shown in Fig. 7.4 (a).

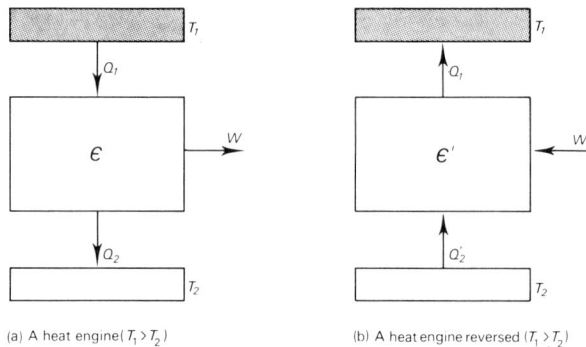

(a) A heat engine ($T_1 > T_2$)   (b) A heat engine reversed ($T_1 > T_2$)

FIG 7.4   A heat engine and a reversed heat engine

Engines (Q and A)

Q.1.   Do you think that an engine in which Q_1 is 100 kW and W is also 100 kW is possible? If you think it is impossible give your reasons for thinking so.

[Impossible, because $\eta = \dfrac{W}{Q_1} = \dfrac{100}{100}$ or 100 per cent]

Q.2.   Reconsider the first question for $Q_1 = -80, W = 80$ kW.

[Possible, because although performance is 100 per cent, the engine is reversed]

Q.3.   Reconsider Question 1 for $Q_1 = 80, Q_2 = -40, W = 80$ kW.

[Impossible. It contravenes first law in that

$Q_1 + Q_2 - W = 80 - 40 - 80 \neq 0$. See equation (6.3)]

Q.4.   An inventor claims he has invented an engine that takes energy from waste liquid in a sewer. A biological process is going on in the liquid that maintains the liquid at a constant temperature while the liquid gives up 0.25 kJ/kg to a heat engine, so acting as a high temperature reservoir for the engine. The flow-rate of the liquid in the sewer is 125 kg/min. The inventor claims that this produces enough energy to supply a 0.5 kW electric fire. Do you believe him?

$$[Q_1 = \frac{125 \times 0.25}{60} = 0.52 \text{ kW}; W = 0.5 \text{ kW}$$

Therefore $Q_1 > W$, and, according to the first and second laws to the extent that we have at present discussed them, his claim is possible]

7.4 Coefficients of performance

In Fig. 7.4 (b) a heat engine is shown reversed. A reversed engine is called a heat pump or refrigerator depending on whether it is being used (a) to deliver energy to a relatively high-temperature reservoir—for example to water used for heating a building—or (b) to take energy from a relatively low-temperature reservoir—for example the contents of the cold chamber of a refrigerator.

The performance of such an arrangement is assessed by what is called its coefficient of performance. Performance in section 6.1 is defined as the ratio

$$\frac{\text{What one gets that is useful}}{\text{What one pays to get it}}$$

In the case of the system in Fig. 7.4 (b) the coefficients of performance are

(a) As a heat pump

$$C_{HP} = \frac{Q_1}{W'} \tag{7.5}$$

(b) As a refrigerator

$$C_{REF} = \frac{Q_2'}{W'} \tag{7.6}$$

$$= \frac{Q_1 - W'}{W'}$$

$$= C_{HP} - 1 \tag{7.7}$$

7.5 An engine's best performance

Let us examine how far we can theoretically improve the efficiency of an engine working between two temperature limits. We have already proved in this chapter that a 100 per cent efficient engine is not a possibility. If an engine cannot have an efficiency of 100 per cent what is the highest efficiency it can have If the absolute limit of efficiency is less than 100 per cent, how much less is it? Is the limit 99 or 90 or 9 per cent? The answer to this question is found by first looking at reversibility.

7.6   A reversible engine

The engine of Fig. 7.4 (a) represents the general case of an engine, taking *in* heat energy Q_1 and giving *out* work energy W. When such an engine is reversed, as shown in Fig. 7.4 (b), and is adjusted to send *out* the same amount of heat energy Q_1 there is no certainty that it will then need to take *in* the same amount of work energy W that the forward-working engine gave out. Suppose the work energy taken in was different and had a value W', without our saying whether W' is larger or smaller than W.

Figure 7.5 (a) on the other hand represents a special case of an engine known as a **Reversible engine.** Such an engine takes *in* the same amount of energy Q_1 and delivers work energy W_r (the suffix r is used here to indicate that the energy transfer is associated with a reversible engine). When such an engine is reversed as shown in Fig. 7.5 (b) and is adjusted to send out the same amount of energy Q_1 as it took in as a forward-working engine, then it takes the same amount of work energy W_r in the forward and reverse directions.

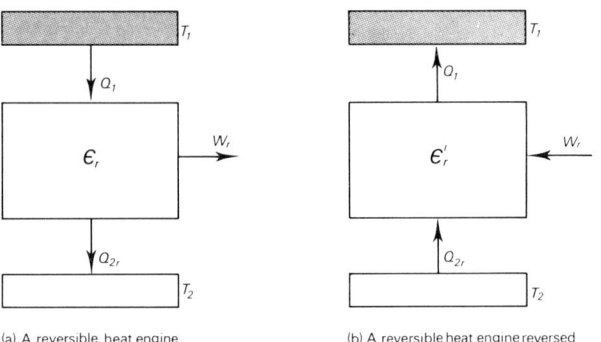

(a) A reversible heat engine   (b) A reversible heat engine reversed

FIG 7.5   A reversible heat engine, operating forward and reversed

From equation (6.8) the efficiency of the reversible engine of Fig. 7.5 (a) is

$$\eta_r = \frac{W_r}{Q_1}$$

(7.)

and its coefficients of performance when reversed as in Fig. 7.5 (b) are

(a) As a heat pump

$$C_{HPr} = \frac{Q_1}{W_r} \qquad (7.9)$$

$$= \frac{1}{\eta_r} \qquad (7.10)$$

(b) As a refrigerator

$$C_{REFr} = \frac{Q_{2r}}{W_r} \qquad (7.11)$$

$$= \frac{Q_1 - W_r}{W_r}$$

$$= C_{HPr} - 1 \qquad (7.12)$$

If we apply the first law in the form of equations (6.3) and (6.5) to the engine forward and reversed of Figs 7.4 (a) and (b) we get

$$Q_1 + Q_2 - W = 0$$
$$Q_1 + Q_2' - W' = 0$$

and, applied to the reversible engine forward and reversed of Figs 7.5 (a) and (b) we get

$$Q_1 + Q_{2r} - W_r = 0$$
$$Q_1 + Q_{2r} - W_r = 0$$

7.7   The best engine

The engines of Fig. 7.5 are special cases in that they are reversible whereas the engines of Fig. 7.4 cover all other cases.

Let us suppose that we have, at any rate in imagination, invented a best forward-working engine that is not reversible but has a greater efficiency than any other engine. Two things follow:

(a) The engine, being irreversible, is in the same category as the engine of Fig. 7.4 (a)
(b) The engine, being more efficient than any other, is more efficient than the engine of Fig. 7.5, and so for the same Q_1, $W > W_r$.

Let this imaginary best forward-working engine drive the reversed-working reversible engine, as shown in Fig. 7.6. Because the imaginary engine is better than the reversible engine $W > W_r$ as stated above and so there will be a positive balance of work $(W + W_r)$ for the combined engine where W is +ve and W_r is −ve.

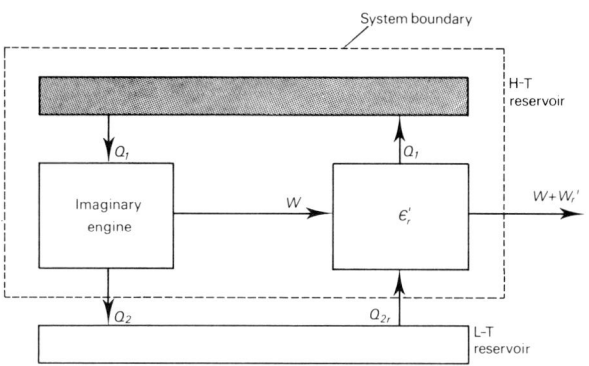

FIG 7.6   Imaginary engine driving a reversible engine reversed

In the arrangement shown in Fig. 7.6 we consider the system within the dotted lines to be cyclic because Q_1 is being transferred both into and out of the high-temperature reservoir. Usually neither the high nor the low-temperature reservoir is cyclic: more likely one is losing energy all the time and the other is gaining it. Here the amount of stored energy in the high-temperature reservoir remains the same because as much heat energy goes in as comes out. Therefore the high-temperature reservoir is working cyclically. The engines are cyclic as stated in Chapter 6, and so everything within the dashed lines in Fig. 7.6 is cyclic as there is no change of energy stored within the system. However, we have here a system that produces positive work energy $W + W_r$ while exchanging heat energy $(Q_2 + Q_{2r})$ with a single energy source. This type of engine the second law states is impossible and therefore it follows that $(W + W_r)$ must be zero or negative and that $W_r \geqslant W$. The imaginary engine therefore is valid only if its performance is equal to or less than that of a reversible engine. We conclude that—assessed by its efficiency—the reversible engine is the best engine and indeed its efficiency is greater than that of any practical engine yet designed. Any future design will be subject to this limitation which can be shown by putting the newly designed engine to the test described above. The efficiency of the reversible engine is the greatest efficiency we can obtain. This is the limitation that applies to all heat engines in which a working fluid undergoes a series of processes that follow a cyclic path giving an overall cyclic process. The working fluid in a heat engine returns to its initial state when it has completed its cycle. It is most important to appreciate this point as well as the facts that the cyclic process has involved no mass transfer or change in chemical composition. If the process were not cyclic there would be a change of stored energy within the system.

How to assess the numerical value of the performance of a reversible engine is discussed in Chapter 12.

7. 8   Summary

A 100 per cent efficient engine is an impossibility. The engine with the greatest possible efficiency of all engines working between energy reservoirs at the same temperatures is a reversible engine. For such an engine its efficiency η is given by

$$\eta = \eta_r = \frac{W_r}{Q_1}$$

and when it is run in reverse its coefficients of performance are given by

$$C_{HPr} = \frac{1}{\eta_r} \quad \text{and} \quad C_{REFr} = \frac{1}{\eta_r} - 1$$

7. 9   Questions for the reader

Q. 1.   An inventor claims to have a reversible engine. When it works as an engine its work energy output is 12 kJ and its efficiency is 0·35. Calculate Q_1 and Q_2 and also find C_{HPr} and C_{REFr}.

[34·3, −22·3 kJ; 2·86, 1·86]

Q. 2.   In Question 1 what are the heat energy flow rates $\dot{Q}_1$ and $\dot{Q}_2$ when the engine is being used as a refrigerator for the two cases when the rate of work energy input is (a) 12 kW and (b) 10 kW?

[34·3, −22·3 kW; 28·6, −18·6 kW]

Q. 3.   A reversible heat pump uses a river, which flows at 80 m³/s, as an energy source, lowering the water temperature by 10°C. What rate of supply of work energy is required to drive the pump and what energy is available for heating if $C_{HPr} = 9$. The specific internal energy of water is 4 kJ/kg K and its density is 1 000 kg/m³ (specific volume 0·001 m³/kg).

[400 MW; 3 600 MW]

Q. 4.   If for a reversible engine the efficiency is measured as $\eta = 0·45$ and its coefficient of performance as a heat pump is $C_{HPr} = 3·27$. Are these consistent?

[No, if $\eta_r = 0·45$, C_{HPr} 2·22, not 3·27]

Q. 5. A newly designed engine which is not reversible claims the following performance parameters. Are they possible?

$$\eta = 0 \cdot 35, \; C_{HP} = 3 \cdot 24$$

[No. C_{HP} must be 2·86 or less]

Q. 6. A refrigerator is part of an air conditioning plant of a building. If the power removed is 15 kW, and it is working on a cyclic process whose coefficient of performance as a heat pump is 8·0, what must be the rating of the compressor?

[2·14 kW]

Q. 7. The system in Question 6 is used in the winter months as a heat pump. If this is so and the system is reversible what power can be put into the building with this system?

[17·14 kW]

Q. 8. Given two reversible engines one working as a forward engine and the other as a reversed engine having one energy source in common. If the forward working engine's efficiency is 0·27 and the reversed engine's coefficient of performance as a heat pump is 3·7 are they

(a) possibly working between the same two thermal reservoirs,
(b) necessarily working between the same two thermal reservoirs?

[Yes, No]

8

Steady-flow energy equation

Not all situations can be analysed using the idea of a system as they have been so far. Here is introduced the concept of a control volume that is like a system in that heat and work may take place at its boundary, but unlike a system in that

(a) *Its shape and size must not change*
(b) *Mass may flow into and out of it across its boundary.*

8.1 A control volume

In Chapters 2 and 3 we discussed the nature of a system, which was described as having a definite and unchanging mass possessing certain properties. Two actions on the boundaries of such systems were described one being heat and the other work which transferred energy Q by heat and energy W by work across the boundaries causing changes to the internal energy U, kinetic energy K, and gravitational energy Z, stored in the system. The relationship between them in any one system during a process was summed up in equation (3.2), derived from the first law

$$Q - W = \Delta U + \Delta K + \Delta Z \qquad (3.2)$$

The symbols Q, W, U, K and Z represent quantities of energy of which the units would normally be kJ, but it might be useful to consider the stored energy of the system in energy per unit mass in the system, the units normally used being kJ/kg. Equation (3.2) would then be written

$$q - w = \Delta u + \Delta k + \Delta z \qquad (8.1)$$

or, if the changes of energy were being considered in terms of rates of change of energy per unit time, equation (3.2) would become

$$\dot{Q} - \dot{W} = \Delta\dot{U} + \Delta\dot{K} + \Delta\dot{Z}$$

In this case the units normally used are kJ/s, i.e. kW. Such a rate of transfer of energy is called **Power**. If the changes of energy were being considered in terms c rates of change of energy per unit mass and per unit time equation (8.1) becomes

$$\dot{q} - \dot{w} = \Delta \dot{u} + \Delta \dot{k} + \Delta \dot{z} \qquad (8.2$$

and the units normally used are $kJ/s/kg$, i.e. kW/kg.

In equations (8.1) and (8.2) the use of lower case symbols signifies that energy per unit mass is being used, and in equation (8.2) the lower case symbols with the dot above signify energy per unit mass per unit time.

Two of the most important things to notice about a heat engine are, firstly, that no mass crosses its boundary and, secondly, that it is cyclic. Engineer are often interested in situations that involve relatively large mass transfers as, for example, air and kerosene entering a gas turbine and exhaust gases—the products of combustion—leaving it (Fig. 8.1). The turbine when working is not ther fore a heat engine, nor can it be regarded as any other sort of system because mass crosses its boundaries. So that we can investigate such a phenomenon we consider a fixed-boundary zone comprising, in this example, the turbine and any auxiliary components and this we call a **Control Volume.** Through the control volume one can imagine a fluid passing while it undergoes a process.

Let us consider steady flows of kerosene and of air entering, and the products of combustion leaving, the control volume of Fig. 8.1. The fluids can be considered as a series of systems passing through the control volume. The contro volume is shown again in Fig. 8.2 where the dashed line shows a mass, or a systen of kerosene entering the control volume. It can be seen from the dashed line, whicl shows the boundary of the system in the first position we consider, that some V_1 o

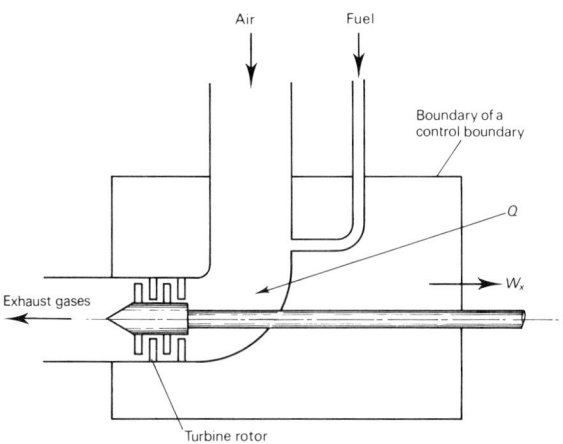

FIG 8.1   A gas turbine inside a control volume

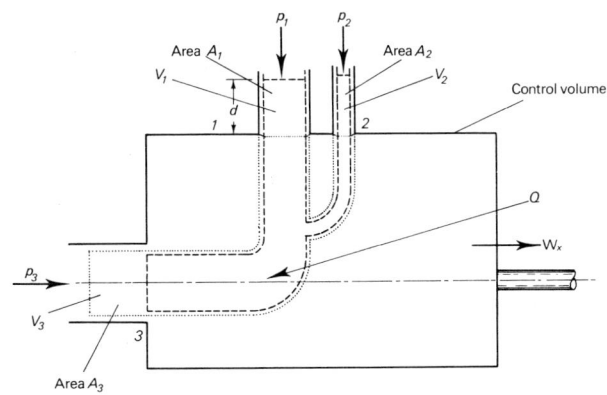

FIG 8.2   A system passing through a control volume

the air at 1, and some V_2 of the kerosene at 2 have not yet entered the control volume, but that none of the exhaust gas at 3 has yet left the control volume. Let a short time pass such that the system moves through the control volume, its second position being shown by the dotted line in Fig. 8.2. In this position the last of the air at 1 and of the kerosene at 2 have entered the control volume and the leading part of the system, V_3 of exhaust gas, has already left the control volume at 3.

The mass of air, kerosene and exhaust gas to which we are referring is considered to be a system and has the properties of a system in that it has a constant mass although its shape and position may change. Any flow of gases through a pipe can in fact be represented by a series of arbitrarily defined systems which follow each other. What we are doing when we consider Fig. 8.2 is to look at one such system in two positions it would occupy at different times.

8.2   External work done from a control volume

In the last section we described a quantity of fluid, a system, passing through a control volume. If the fluid involved in this process is in a condition of what is called steady flow, that is to say if the kerosene and air are *entering* the control volume in Fig. 8.2 at steady rates $\dot{M}_{Kero}$ and $\dot{M}_{Air}$ units of mass per unit time and if the exhaust gases are *leaving* at a steady rate $\dot{M}_{Exh}$, then

$$\dot{M}_{Kero} + \dot{M}_{Air} - \dot{M}_{Exh} = 0 \qquad (8.3)$$

In conditions of steady flow the mass of fluid entering the control volume equals the mass leaving it so that there is no change of the mass of fluid actually within the control volume.

Consider the work energy transfers that take place at the boundary of the control volume in the short interval of time during which the system passes from the first (dashed) position shown in Fig. 8.2 to the second (dotted) position.

There are four work transfers:

(a) Work transfer by the surroundings to the air of $p_1 A_1 d_1$ (force $p_1 A_1 \times$ distance d_1), or $p_1 V_1$ (because volume $V_1 = A_1 d_1$) units of energy. This can be thought of as the work required to push the last of the air of the system into the control volume by the first of the air of the next system.

(b) Work transfer by the surroundings to the kerosene of $p_2 V_2$ units of energy. The last of the kerosene entering the control volume.

(c) Work transfer of $p_3 V_3$ units of energy is done by the exhaust gas in displacing the surroundings. The first of the system is pushing its way out of the control volume.

(d) Useful work transfer by the system on the surroundings of work energy W_x transmitted by the turbine rotor and shaft, see Fig. 8.1

In this relatively simple state of affairs of a system passing through an idealised gas turbine inside a control volume these four work transfers together make up the work energy W done by the system, so that

$$W = W_x + p_3 V_3 - p_1 V_1 - p_2 V_2$$

$$= W_x + \sum pV \qquad (8.4)$$

The other action occurring at the boundary is that of heat, the heat energy Q entering the control volume and also the system. We can use this fact and equations (8.4) and (3.2) to write down an energy balance for the control volume through which a series of such systems is passing in steady flow

$$Q - W_x - \sum pV = \Delta U + \Delta K + \Delta Z$$

or

$$Q - W_x = \Delta U + \sum pV + \Delta K + \Delta Z$$

$$= \Delta [U + pV] + \Delta K + \Delta Z$$

Because the processes of getting air, kerosene and exhaust gases into and out of the control volume are constant-pressure processes we can put

$$H = U + pV \qquad (8.5)$$

and then $$Q - W_x = \Delta H + \Delta K + \Delta Z \qquad (8.6)$$

H is called **Enthalpy.** In order to express a change of enthalpy equation (8.5) must be differentiated,

$$\Delta H = \Delta U + p\Delta V + V\Delta p \qquad (8.7)$$

8.3 Enthalpy

Hitherto we have dealt only with changes of stored energy. That is to say we have concerned ourselves with ΔU, ΔK and ΔZ, not with absolute values of U, K and Z. Our reason for discussing only changes is that we have not considered what it means for a substance to have absolute zero energy. It is sufficient for our purpose at present to consider what is our view of the mass in a system if there were no stored energy. As has been already stated in section 3.5, if there were no stored energy there would be no relative movement of molecules, no forces between molecules, and no forces between atoms, and also no forces between electrons, protons and neutrons within atoms. This is not a feasible state and so the lowest amount of stored energy that we can imagine is the stored energy the system would have at absolute zero temperature. Because of the difficulty of observing matter at absolute zero temperature the reader is asked to picture the atoms of matter in this state as motionless inert grains of sand, which if they could be packed closely together would occupy a very small volume.

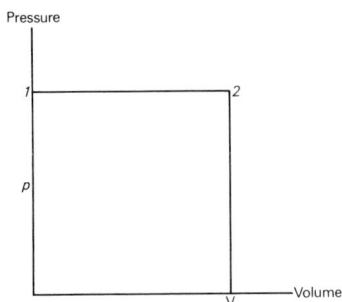

FIG 8.3   Change of state from zero energy

In Fig. 8.3 state 1 of a system, expressed in terms of pressure and volume, is that of a system with effectively zero energy and therefore zero volume, in surroundings at constant pressure p. Consider the system to have energy Q transferred to it by heat; then the previously inert particles would be excited by the energy so that the system would start occupying a finite volume. In this case, if the system generally remains with its centre of gravity near the same place in space in going from state 1 to state 2,

$$\underset{1-2}{\Delta} Z = 0 \tag{8.8}$$

also $$\underset{1-2}{\Delta} K = 0 \tag{8.9}$$

Because the change of internal energy commenced at zero the value of ΔU from state 1 to state 2 is equal to the absolute value of U at state 2. That is to say

$$\underset{1-2}{\Delta} U = U - 0$$

$$= U \qquad (8.10$$

If the volume of the system changes from zero in state 1 to V in state 2 the work done is

$$\underset{1-2}{W} = p(V - 0)$$

$$= pV \qquad (8.11$$

Because the change has in fact been made at constant pressure. Substituting in equation (3.2) all the values of energy changes given in (8.8), (8.9), (8.10) and (8.11), we get

$$\underset{1-2}{Q} - pV = U$$

$$\underset{1-2}{Q} = U + pV$$

or from equation (8.5)

$$H = U + pV$$

$$= \underset{1-2}{Q} \qquad (8.12$$

This is consistent with equation (8.7) because this change from state 1 to 2 is carried out at constant pressure and therefore $V \Delta p = 0$. Because, at absolute zero, both U and V are zero, H also is zero at this temperature. From this discussion it should be clear that enthalpy may be considered as the gain in stored energy of a system in going from state 1 to state 2 in the process proposed in Fig. 8.3. The **Enthalpy** of a system is the stored energy equal in amount to the energy that would have to be transferred to the system, if the only means of inward transfer were by heat at constant pressure, to bring the system from a state of zero energy to its present state.

There may appear to be a contradiction between the statement made above defining enthalpy, particularly in the words 'at constant pressure', and the term $V \Delta p$ in equation (8.7). There is no contradiction for the following reasons. It was stated as a comment on equation (8.12) that because, at absolute zero temperature, both U and V are zero H also has a value zero. H_2 at point 2 in Fig. 8.3 has therefore an absolute value. If the reader turns to Fig. 8.4 he will see that H can have one value at point 2 with pressure p and volume V but a different value at point 3 where the pressure is $p + \Delta p$ and volume $V + \Delta V$. The ΔH of equation (8.7 is $(H_3 - H_2)$ and the Δp and ΔV are differences in these variables between states 2 and 3.

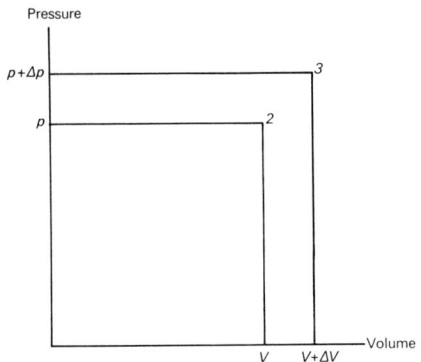

FIG 8.4   Change of state from 2 to 3

8.4 The energy equations

The steady-flow energy equation is applied to a control volume, where mass crosses the boundary steadily into and out of the volume so that the mass inside the volume remains constant. It can be written in the form of equation (8.6)

$$Q - W_x = \Delta H + \Delta K + \Delta Z$$

whose terms in this form have units of energy. In terms of change of energy per unit mass passing through the control volume it is often written as

$$q - w_x = \Delta h + \Delta k + \Delta z \tag{8.13}$$

or per unit mass per unit time

$$\dot{q} - \dot{w}_x = \Delta \dot{h} + \Delta \dot{k} + \Delta \dot{z} \tag{8.14}$$

These are steady-flow energy equations, meaning that the fluid flow-rates into and out of the control volume to which they refer are equal and constant. Earlier we dealt with non-flow energy equations (3.2), (8.1) and (8.2). In equations (3.2) and (8.6); (8.1) and (8.13); (8.2) and (8.14) the units normally used are kJ, kJ/kg and kW/kg respectively.

In equation (3.3) it was stated that a change ΔU of the internal energy when the temperature of a mass M of a fluid is changed from T_1 to T_2 is given by

$$\Delta U = Mc_v(T_2 - T_1) \tag{8.15}$$

Similarly the change in the enthalpy is given by

$$\Delta H = Mc_p(T_2 - T_1) \tag{8.16}$$

where c_V is called the specific internal energy per degree, and c_p the specific enthalpy per degree. The definitions of c_V and c_p for a perfect gas are explained more fully and defined in Chapter 18.

Steady-flow energy equation (Q and A)

Q. Water from a constant-head tank runs out of the tank through a hole in the bottom into a vertical pipe with smooth walls that are thermally insulated from the surroundings. The pipe is open at the bottom end. If the constant head of water above the open end of the pipe is 15 m find the velocity of water leaving the open end (Fig. 8.5). The constant head is maintained by the flow of water being controlled by a ball cock. Assume the temperature remains constant.

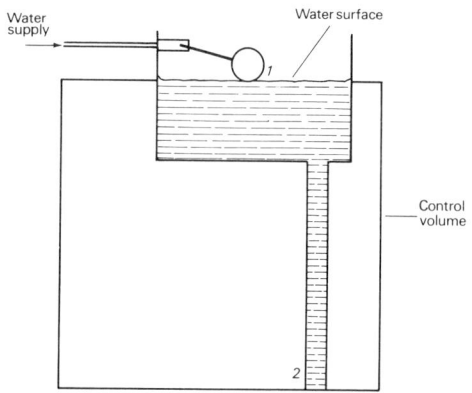

FIG 8.5   To illustrate question on steady-flow energy equation

A. Enclose the apparatus in a control volume.

There is no heat, $\therefore Q = 0\ kJ$
There is no work, $\therefore W_X = 0\ kJ$

There is no change of temperature and pressure, $\therefore \Delta H = 0\ kJ$ (equation (8.16)). Let the water surface in the tank be at 1, and the open end of the pipe be at 2.

Assume $V_1 = 0$ and $\bar{z}_1 = 15$ m. We want to find the value of V_2 if $\bar{z}_2 = 0$. The change of kinetic energy of water is given by

$$\Delta K = M\ \frac{(V_2^2 - 0^2)}{2}\ kg(m/s)^2$$

$$= \frac{MV_2^2}{2}\ kg(m/s)^2$$

$$= \frac{MV_2^2}{2} \; \text{N m}$$

$$= \frac{MV_2^2}{2 \times 10^3} \; \text{kJ}$$

The change of potential energy of water is

$$\Delta Z = M \; 9 \cdot 81 \; (0 - 15) \; \text{kg}(\text{m/s})^2$$

$$= - \; M \; 147 \; \text{N m}$$

$$= - \; M \frac{147}{10^3} \; \text{kJ}$$

Substituting for these in equation (8.6)

$$Q - W_X = \Delta H + \Delta K + \Delta Z$$

$$0 - 0 = 0 + \frac{MV_2^2}{2 \times 10^3} - \frac{147M}{10^3}$$

$$V_2 = 17 \cdot 2 \; \text{m/s}$$

Note about units

In the answer to the question we have just considered, ΔZ is given in units of $\text{kg}(\text{m/s})^2$ and also in joules. Every term must be in the same units and it appears that here there is an inconsistency of units in using J and $\text{kg}(\text{m/s})^2$ as interchangeable units. However, Newton's second law states that

force = mass × acceleration

or $N = \text{kg m/s}^2$

Hence $\text{kg}(\text{m/s})^2 = \text{kg}(\text{m/s}^2)\text{m} = \text{Nm}$. By definition, one N m is one J.
Therefore it follows that J has the dimensions $\text{kg}(\text{m/s})^2$. If Q, W_X and ΔH are in terms of kJ and not J then ΔK and ΔZ must be divided by 10^3 to reduce them to the same kJ units.

8.5 Summary

Control volumes, across the boundaries of which mass flows, have been described, and a flow of fluid through the control volume has been studied. The flowing fluid has been considered as a series of systems, one following another through the control volume. The first law, as described in equation (3.2), is applied to a system passing through the control volume under steady flow conditions, and the first law applied to the control volume is derived and is known as the steady-flow energy equation. Enthalpy was defined and compared with internal energy, and the difference between them was described in terms of the state of a system and its relationship to the state of the system when storing no energy.

8.6 Questions for the reader

Q. 1.   4 kg/s of air at a pressure of 13×10^5 N/m^2 and a temperature of 800°C enters an adiabatic turbine through a pipe of 100 mm internal diameter and leaves it through a 300 mm pipe. On leaving the turbine the air pressure and temperatur$_e$ are 1×10^5 N/m^2 and 550°C respectively. How much work energy per kg does the air do inside the turbine? The specific volume of air at temperatures and pressures 1073 K, 13×10^5 N/m^2 is 0·237 kJ/kg. 823 K, 1×10^5 N/m^2 is 0·236 m^3/kg. The specific enthalpy per degree K of air at constant pressure is 1·01 kJ/kg K.

[253 kJ/kg]

Q. 2.   20 kg/s of water initially with negligible velocity, are flowing down a water-fall 20 m high. During the fall the water gains only 75 per cent of the velocity it would have gained if falling freely in the Earth's gravitational field, because of friction between the river bed and the water. Assume that the river bed and the atmosphere are both non-conductors, and calculate the velocity of the water at the foot of the falls. c_v for the water is 4 kJ/kg K.

[14·9 m/s]

Q. 3.   Is there a difference in temperature of the water between water at the top o$_f$ the fall in Question 2 and the same water streaming away at the bottom? If so how$_-$ much?

[0·0213°C rise]

Q. 4.   Assume that water of Question 2 falls into a deep wide basin at the bottom. Now does the temperature of the water change? If so, by how much?

[0·049°C rise]

Q. 5.   A fault has arisen in a tooth paste filling machine. The tops of the tubes have not been capped. Each tube is filled from the bottom at the rate of 10^{-2} m^3/s—the top of the tube has a 50 mm diameter hole. Calculate the speed at which the tooth paste is coming out, assuming the worst possible case.

[5·1 m/s]

Q. 6.   Following on from Question 5, the production manager's desk is located 10 m below the top of the discharging tube. If the tooth paste was originally at 50°C at what temperature is it after landing on the desk? Assume c_v for tooth paste is 3·00 kJ/kg K.

[T = 50·037°C]

Q. 7. An engine takes in 7·5 g/s of air at 25°C which, by reacting with petrol in the engine, releases 15 kW of power. If 6 kW of this is transmitted as work energy out of the system, and $5^{1}/_{2}$ kW is transmitted to the cooling system, what is the temperature of the exhaust gases? Assume the mass of fuel is negligible as are the inlet and outlet velocities of air and exhaust gases respectively. Assume $c_p = 1·00$ kJ/kg K for the air and the exhaust gases.

[492°C]

Q. 8. If the outlet velocity of the exhaust gases were not negligible and the outlet temperature of the gas were found to be 485°C what would be the exhaust velocity?

[118 m/s]

9

Applications of the steady-flow energy equation

The steady-flow energy equation,

$$Q - W_x = \Delta H + \Delta K + \Delta Z$$

derived in the last chapter, is considered in various modified forms and these are applied to particular cases in many of which it is seen that the equation may be used in a reduced form.

9.1 Steady flow

Steady flow is achieved when the mass flow-rate and the energy exchange rates reach a constant value. A general example of fluid flow is seen in Fig. 9.1. The fluid in the pipe is seen passing steadily through a control volume exchanging heat energy Q with its surroundings perhaps by conduction through the walls and exchanging work energy W_x perhaps by means of a rotor on a spindle

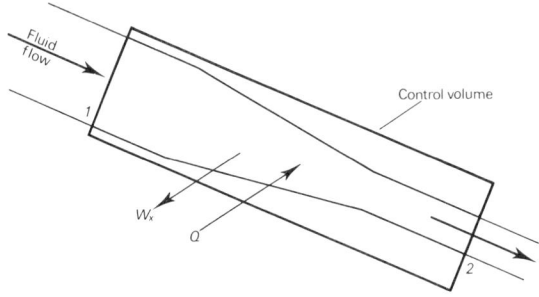

FIG 9.1   Fluid flow in a pipe

projecting through the pipe wall into the stream of fluid. In terms of unit mass flow cf the fluid these energy exchanges will be q and w_x. Also if the pipe is not level and the diameter varies the gravitational and kinetic energies will change by Δz and Δk per unit mass of fluid. In the last chapter we stated that the relationship between these transfers and changes of energy is expressed by equation (8.13)

$$q - w_x = \Delta h + \Delta k + \Delta z$$

By substituting for Δk and Δz the values from equations (3.4) and (3.5), this becomes

$$q - w_x = (h_2 - h_1) + \frac{V_2^2 - V_1^2}{2} + (\bar{z}_2 - \bar{z}_1)g \qquad (9.1)$$

9.2 Steady flow when Q and W_x are both zero

There is a very simple case shown in Fig. 9.2 in which the flow is adiabatic because the pipe is thermally insulated and therefore q = 0, and because the pipe is electrically insulated and rigid (also there is no rotor) $w_x = 0$. The pipe is level making $\Delta z = 0$, and there is no change in velocity so $\Delta k = 0$, hence

$$0 = \Delta h$$

or $h = \text{constant}$ $\qquad\qquad\qquad\qquad\qquad\qquad\qquad\qquad (9.2)$

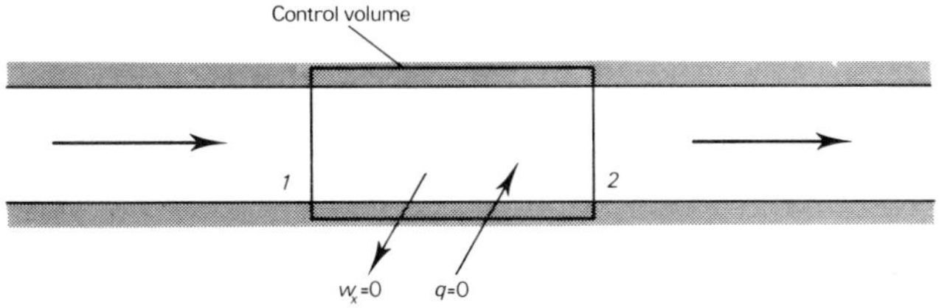

FIG 9.2   Flow in an insulated rigid pipe

Equation (9.2) is an important statement because there are many quite complicated cases in which changes of velocity and level are relatively so small that it is sufficiently accurate to assume Δk and Δz are negligible and that enthalpy is constant. In these cases it is sufficiently accurate to use (9.2) as a shortened version of the steady-flow energy equation. To decide whether Δk and Δz are small enough to be neglected each case must be examined and a decision made.

In those cases in which q and w_x are zero and there is a sufficient change of level to make Δz significant in the equation but changes of velocity make Δk negligible, then equation (9.1) becomes

$$0 = h_2 - h_1 + (\bar{z}_2 - \bar{z}_1)g \qquad (9.3)$$

and if the change of level is insignificant but there are significant changes of velocity, then equation (9.1) becomes

$$0 = h_2 - h_1 + \frac{V_2^2 - V_1^2}{2}$$

or

$$h_2 + \frac{V_2^2}{2} = h_1 + \frac{V_1^2}{2}$$

that is

$$h + \frac{V^2}{2} = \text{constant}$$

(9.

In equation (9.4) it is seen that the enthalpy h of the fluid at any point is less by an amount $\frac{1}{2}V^2$ than it would be if the fluid were at rest. The enthalpy that the fluid would have when stationary is called the **Stagnation enthalpy** h'. Its value is given by

$$h' = h + \frac{V^2}{2}$$

If there had been a restriction in the pipe the case would have looked something like that shown in Fig. 9.3. The restriction could have been intentional as by a control valve, or accidental as by an unwanted chemical deposit. In the case shown in Fig. 9.3, q and w_x are both zero. A restriction of this kind is usually associated with a drop in pressure downstream. That is to say the pressure in the neighbourhood of 2 is lower than that at 1. If the fluid were a gas an increase in specific volume of the gas would result from the lower pressure at 2 leading to an increased velocity. To such a case equation (9.4) would apply. Velocity at 2 for any particular velocity at 1 depends also on pipe dimensions. For large diameter pipes, velocity and kinetic energy changes can usually be neglected.

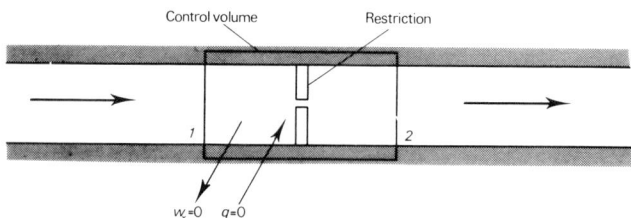

FIG 9.3   Flow in an insulated rigid pipe with a restriction

There are times when fluid, in the form of a vapour (a fluid in the region between behaving as a liquid and behaving as a gas) is passed through a convergent-divergent nozzle as in Fig. 9.4 with the intention of reducing its pressure in order to reduce its saturation temperature. Such a nozzle used in this way is called a throttle. A throttle is used for this purpose in a vapour refrigerator (see Chapters 11 and 26). There is no difference in principle between such a throttle—shown in Fig. 9.4 and the restriction shown in Fig. 9.3. In the throttle of Fig. 9.4 the diameters of the pipe at entry and discharge are the same and the reduction of pressure often leads to an increase of specific volume and consequently an increase of velocity. However, it is normal in such refrigerators for the change in kinetic energy across the throttle to be small enough to be treated as negligible and for the form of the steady flow energy equation of section 9.2 to be used.

Restrictions of the kind shown in Fig. 9.4 and its associated pipework can be designed to make the change of kinetic energy either negligible or to make it significant as in Fig. 9.5. The restriction of the nozzle in Fig. 9.5 is convergent and intentionally increases the velocity of the fluid for use, in, say, a turbine where the high velocity fluid leaving the nozzle at 2 is made to strike blades on the rotor of a turbine and so to turn the turbine shaft. While the fluid passes

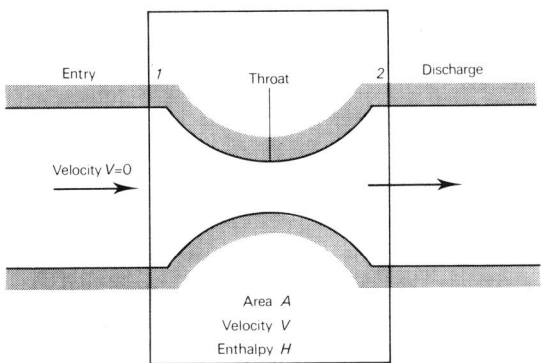

FIG 9.4   Flow through a convergent-divergent nozzle

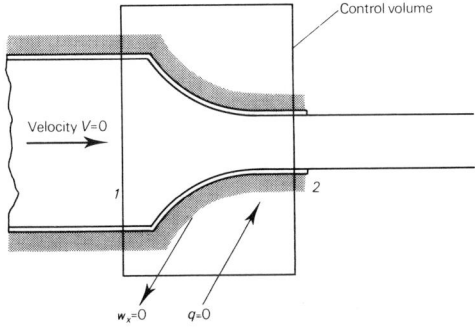

FIG 9.5   Flow through a nozzle

through the control volume some of its enthalpy is converted to kinetic energy—
therefore here equation (9.4) applies.

A more complicated situation is shown in Fig. 9.6 which is that of an
insulated rigid level pipe without any moving parts other than the fluid and therefore q and w_x both equal zero. The fluid is a mixture of reactants passing across
the control volume boundary at 1. If the equipment is properly designed the mixture's temperature is raised by the energy released due to the combustion proces
as the fluid passes through the control volume. What is happening within the fluid
mixture during a chemical reaction is that some bonds between atoms are re-
arranged causing a change in the energy used in binding the atoms into molecules.

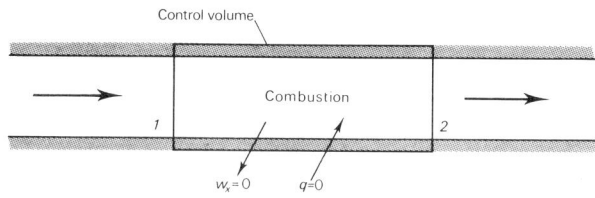

FIG 9.6   Combustion taking place in a steady flow

A reaction in which more energy is released in bonds breaking than is taken up in
the making of new bonds is called **EXOTHERMIC**. The nett energy released in an
exothermic reaction increases the temperature of the new gases formed, and hence
their specific volume is greater on leaving than that of the reactants entering. The
increase in volume of fluid in the control volume, in this case a combustion chamber increases the kinetic energy of the fluid which could be used to drive a jet
engine in a direction opposite to the direction of flow. The form of steady flow
energy equation that applies in this case is equation (9.4) as the energy released b
combustion is a part of the energy included in the enthalpy h (see Chapter 20).

9.3 Steady flow when Q is zero but W_x is not

In section 9.2 all the cases discussed were examples of the application of the steady flow energy equation to cases in which q and w_x were zero. In
this section the cases discussed have in common their being applications of the
same equation where q is zero. The difference between these and those of section
9.2 is that in this section w_x is not zero.

The first case is shown in Fig. 9.7 and is an example of a water-
driven turbine. There is a work output from the turbine that is derived from a loss
of gravitational energy by water in a hydroelectric plant. If the turbine and the
pipework within the control volume are well insulated there is no heat, the process
undergone by the fluid will be adiabatic, so q will be zero. If the water temperatures at points 1 and 2 are not significantly different $T_2 = T_1$ and $\Delta h = 0$. Then if

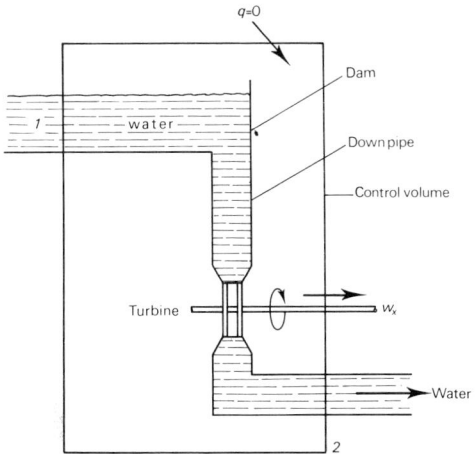

FIG 9.7   Water Turbine

the water velocity at 1 is negligible equation (9.1) becomes

$$- w_x = \frac{(V_2^2 - 0)}{2} + (\bar{z}_2 - \bar{z}_1)g$$

or
$$w_x = - \frac{V_2^2}{2} - (\bar{z}_2 - \bar{z}_1)g \qquad (9.5)$$

The negative sign before $\frac{1}{2}V_2^2$ in equation (9.5) shows that the larger is V_2 the smaller is the useful work w_x.

    In the case of the impulse gas turbine in Fig. 9.8 consider unit mass of the gas entering, at A, the space between adjacent stationary blades, SB. Its flow is diverted by the blades to a direction B making, as it enters the control volume, an angle α with the direction D of motion of the moving blade MB. The moving blades have leading edge angles β relative to direction D. The fluid in passing through the moving blades exchanges energy with them. The moving blades turn the direction of flow of the gas relative to themselves from a direction making an angle β to another making an angle $(180-\beta)$ with the direction D. This change of direction of the flow causes a change of momentum of the gas and causes a force to act on the moving blades. The moving blades are therby driven around the polar axis of the turbine, so driving the rotor and the shaft, which does work w_x units of energy per unit mass of gas flowing.

    There are four directions of flow of gas to consider:

    (a)   The direction α of the gas leaving the stationary blades. This can be seen in Fig. 9.8 and is the same angle as that of the stationary blades' trailing edge, this being the direction of motion of the gas as it enters the control volume.

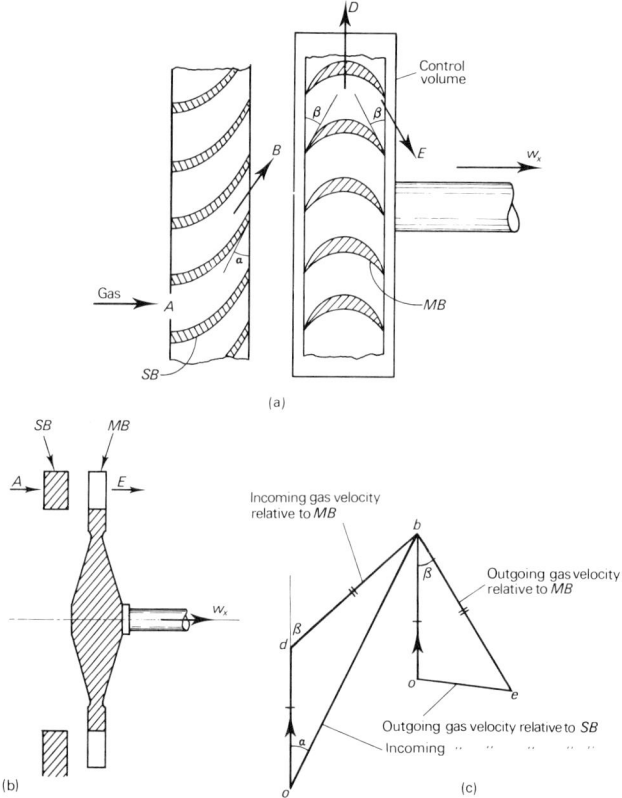

FIG 9.8   Flow through part of a turbine (a) Blades and gas flow
(b) Blades, rotor and shaft (c) Velocity diagram for gas

(b)   The direction β of the gas immediately after entering the control
      volume. If conditions are ideal the direction of the gas does not
      change between (a) and (b), the angles α and β being different be-
      cause α is associated with a stationary blade and β is associated
      with a moving blade.

(c)   The direction $(180-\beta)$ of the gas immediately before leaving the
      spaces between the moving blades. The gas has done work in
      moving the rotor and is now leaving the blades at nearly the
      same velocity, hence with nearly the same kinetic energy, as it
      had on entering.

(d)   The direction of flow of the gas on leaving relative to the station-
      ary objects. This is related to the gas in (c) as (a) is related to (b

Figure 9.8 (b) shows another view of the moving blades relative to the stationary blades.

Figure 9.8 (c) shows a conventional velocity diagram for the blades and the gas. The point o represents the stationary state, od the linear blade velocity in direction D relative to a stationary point. The lines ob and db are drawn in directions making angles α and β respectively with direction D, these angles having known values. The lengths ob and db are proportional to the velocities of the gas relative to the stationary and moving blades respectively. The direction ob is the same as direction B. A line ob may then be drawn also representing the blade velocity making ob parallel to and equal to od. The line be, equal in length to db, can then be drawn at an angle β to the direction of ob because one of the design features of an impulse turbine is that the velocity of the gas relative to the moving blades remains constant. The lengths of the lines be and oe are proportional to the velocities as the gas leaves the moving blades relative to the moving blades and relative to stationary objects respectively.

The flow of fluid through a turbine can be regarded as an example of adiabatic steady flow where useful work energy w_X is not zero. In fact the reason one uses a turbine is because it has ability to change enthalpy in a fluid to work energy. If we consider the whole of a turbine it is usually convenient to treat the process within the control volume as adiabatic and so state that $q = 0$. In practice a turbine always has an energy transfer by heat to the surroundings but in this analysis, by treating the turbine as adiabatic, such losses are considered either to be negligibly small or to occur in some other part of the equipment such as the cooler or condenser into which the turbine discharges.

Figure 9.9 shows schematically an adiabatic turbine doing work w_{xt} by means of a rotating shaft that carries the work energy out of the control volume. The gas or vapour that does the work in the turbine enters the control volume at point 2 and leaves it at 3 after passing through the turbine—which internally might

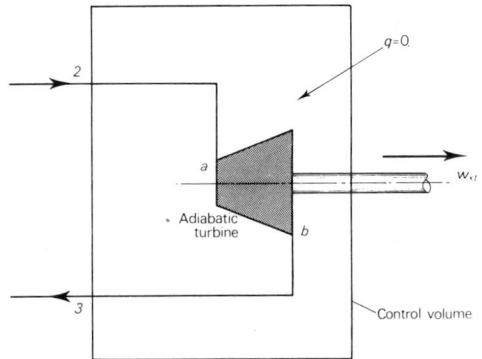

FIG 9.9   A turbine within a control volume

consist of a series of pairs of stationary and moving blades such as the pair shown in Fig. 9.8 (a). Because the turbine is adiabatic q = 0, and the steady flow energy equation (8.13) can be written

$$0 - w_{xt} = \Delta h + \Delta k + \Delta z$$

Usually in the case of a gas or vapour $\Delta z = 0$ and the equation becomes

$$w_{xt} = -\Delta h - \Delta k \qquad (9.6)$$

and sometimes, more often in vapour than gas turbines, $\Delta k = 0$, reducing the equation further to

$$w_{xt} = -\Delta h$$
$$= -(h_3 - h_2) \qquad (9.7)$$

Figure 9.10 shows an adiabatic compressor receiving work energy w_{xp} by means of a rotating shaft that carries work energy into the control volume.

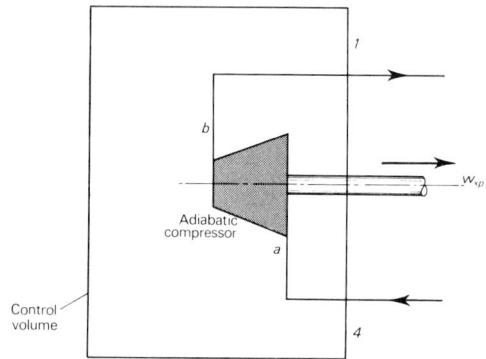

FIG 9.10   A compressor within a control volume

The gas, vapour or liquid that has to be compressed enters the control volume at a relatively low pressure at point 4 and leaves it at a higher pressure at point 1 after passing through the compressor. The fluid may, inside the control volume, pass through one or more cylinders when it is compressed by a reciprocating pump. Such a pump is particularly suitable for so-called incompressible fluids— liquids which have only very small volume changes for very large pressure changes, for example saturated water at 0.03×10^5 N/m² has a volume only 1·4 times that of saturated water at 75×10^5 N/m². When the fluid to be compressed is a vapour or a gas the compressor is likely to be similar to a reversed turbine or to be a fan. Because the compressor is adiabatic, q = 0, and the steady flow energy equation can be written as equation (9.6)

$$w_{xp} = -\Delta h - \Delta k \qquad (9.8)$$

w_{xp} has been substituted for w_{xt} as we are now talking about a pump. When a liquid is being compressed the value of Δk is usually negligible and the form used is that of equation (9.7)

$$w_{xp} = -\Delta h$$
$$= -(h_1 - h_4) \qquad (9.9)$$

It is of course necessary to consider each case when deciding what terms of equation (8.13) can be assumed negligible. In general w_{xp} in the case of a compressor where the working fluid is a liquid is small compared with w_{xp} for the same pressure ratio where the working fluid is a gas.

9.4 Steady flow when W_x is zero but Q is not

Heat exchangers such as those shown in Fig.6.3 might be boilers, condensers, water coolers, the cold chamber of a refrigeration plant or an evaporator. One tube of such a heat exchanger is shown in Fig.9.11 where a fluid is seen flowing through a simple pipe. The difference between this and other cases mentioned in this chapter is that the walls of the pipe are not thermally insulated. In fact, generally, such a pipe is made from a material that is a good conductor because the purpose of a heat exchanger is to transfer energy q from a fluid that is, say, the high-temperature products of combustion that pass across the outer surface of the tube to another fluid inside the tube—this fluid's temperature will rise.

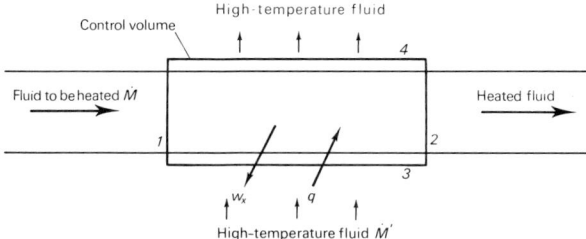

FIG 9.11   Flow in a fluid Heat-exchanger tube

Considering the control volume of Fig.9.11 applied only to the cold fluid, because the walls are rigid $w_x = 0$, and if the change of level were insignificant, as it would be in a heat exchanger, $\Delta z = 0$. Equation (8.13) would become

$$q = \Delta h + \Delta k \qquad (9.10)$$

and if the specific volume and so the velocity of flow were not significantly increased, $\Delta k = 0$, and the equation further reduces to

$$q = \Delta h$$

$$= h_2 - h_1 \tag{9.11}$$

Of course a similar equation can be written for the hot fluid, the flow of which is across the outer surface of the pipe, entering the control volume at 3 and leaving it at 4. For this fluid the equation equivalent to (9.11) would be

$$q' = h_4 - h_3 \tag{9.12}$$

If the flow-rate of the cold fluid were $\dot{M}$ units of mass per unit time and of the hot fluid were $\dot{M}'$ units of mass per unit time, then equations (9.11) and (9.12) may be written in terms of power as

$$\dot{Q} = \dot{M}(h_2 - h_1) \tag{9.13}$$

and $$\dot{Q}' = \dot{M}'(h_4 - h_3) \tag{9.14}$$

As the equations represent energy being transferred out of one fluid into another, if we assume no energy transfer to the surroundings, then

$$\dot{M}(h_2 - h_1) + \dot{M}'(h_4 - h_3) = 0 \tag{9.15}$$

Steady flow (Q and A)

Q.1. There is a storage tank at the top of a house to which sufficient water is supplied to keep the surface of the water in the tank at a constant level of 1 m above the tank bottom. Water is led by a well-lagged (thermally insulated) 20 mm internal diameter pipe from the bottom of the tank to a 10 mm nozzle from which it is discharged horizontally, 10 m below the water surface. If friction is negligible, find the velocity of flow from the nozzle.

A.1. As the constant height of the free surface of the water above the nozzle is given the depth of the water in the tank is irrelevant. The change in temperature of the water is zero because $q = 0$, and there is no friction—therefore $\Delta h = 0$. Assuming the pipe walls are rigid, then $w_x = 0$. As the pipe is well-lagged $q = 0$. From equations (8.13) and (9.1)

$$0 = \Delta k + \Delta z$$

$$= \frac{V_2^2 - V_1^2}{2} + (\bar{z}_2 - \bar{z}_1)g$$

$$= \frac{V_2^2}{2} - 10 \times 9 \cdot 807$$

$$V_2 = 14 \text{ m/s}$$

Q.2. The water level in a reservoir is kept constant. Water drains from the bottom of the reservoir on to turbines and then the water discharges into a river. The velocity of flow of the water from the discharge pipe of the turbine is 10 m/s and located 100 m below the reservoir surface. Find how much work is done by the water in the turbine per unit mass of water. Assume all processes are frictionless and adiabatic.

A.2. Because the process is adiabatic q = 0, and if the kinetic energy of the water at the surface is zero:

$$\Delta k = \frac{V_2^2 - V_1^2}{2} = \frac{10^2}{2} = 50 \text{ Nm/kg} = 50 \text{ J/kg}$$

Assume no change in temperature, and so

$$\Delta h = 0$$

The water falls 100 m, and therefore

$$\Delta z = 9 \cdot 81 \times 100 = 981 \text{ J}$$

So the steady flow energy equation (8.13) becomes

$$0 - w_x = 0 + 50 - 981$$
$$w_x = 931 \text{ J}$$

Q.3. Find the work done per unit mass of working fluid by an adiabatic gas turbine such as that in Fig. 9.9 which admits 1 kg/s of gas at 1 200 K with density 0·2941 kg/m³ (specific volume 3·400 m³/kg) through a 200 mm internal diameter pipe (point 2 in Fig. 9.9) and discharges it through a 100 mm pipe at 300 K and density 1·177 kg/m³ (specific volume 0·85 m³/kg). The specific enthalpy per degree, c_p, for air may be assumed constant at 1·17 kJ/kg K.

A.3. q = 0 because the system is adiabatic

$$\Delta h = 1 \times 1 \cdot 17 (300 - 1\,200) \text{ see equation (8.16)}$$
$$= - 1\,053 \text{ kJ/kg}$$

$$V_3 = \frac{4 \times 0 \cdot 85}{\pi \times 0 \cdot 1 \times 0 \cdot 1} = \frac{340}{\pi} = 109 \text{ m/s}$$

$$V_2 = \frac{4 \times 3 \cdot 4}{\pi \times 0 \cdot 2 \times 0 \cdot 2} = \frac{340}{\pi} = 109 \text{ m/s}$$

assuming the fall is negligible

$$\Delta z = 0$$

So the steady flow energy equation reduces to

$$-w_x = -1\,053$$

or $$w_x = 1\,053 \text{ kJ/kg}$$

9.5   Summary
The application of the steady flow energy equation to cases where, (a) neither heat nor work is done, (b) work but not heat is done and (c) heat but not work is done. The circumstances studied include fluid flow in a sloping uniform pipe, in a rigid insulated pipe, through a throttle, a nozzle, when combustion occurs, through a turbine, turbine blades and a tube from a heat exchanger.

9.6   Questions for the reader
Q. 1.   A perfect gas is flowing with negligible velocity along a thermally and electrically insulated smooth rigid level pipe. Does the gas change its state during the flow and if so, in what respect does it?

[No change]

Q. 2.   A restriction in an insulated rigid level pipe causes a gas that is flowing steadily through it to change its pressure by 1×10^5 N/m^2. During its passage through the restriction what changes of state would you expect in the gas?

$$[-\Delta h, \ +\Delta k, \ \Delta z = 0]$$

Q. 3.   The change of pressure causes the velocity of the gas to increase from being negligible to 125 m/s. Evaluate Δh, Δk in terms of energy per unit mass and evaluate ΔT. The specific enthalpy per degree of the gas is 2 kJ/kg.

$$[\Delta h = -\,7{\cdot}81; \ \Delta k = +\,7{\cdot}81 \text{ kJ/kg}; \ \Delta T = -\,3{\cdot}91 \text{ K}]$$

Q. 4.   In a heat exchanger 10 kg/s of propane and 150 kg/s of air, both at 25°C, enter a uniform rigid combustion chamber and the products of combustion, after burning, leave at 100°C. Across the outer surfaces of the combustion chamber water flows at an initial temperature of 20°C and leaves at a final temperature of 100°C. Find the flow-rate of the water. The specific enthalpy per degree of the gases and of the products of combustion are 1 kJ/kg and of water is 4 kJ/kg K respectively. On combustion 1 kg of propane produces an energy release of 50 MJ.

[1 525 kg/s]

Q. 5.   The propane and air mentioned in Question 4 enter the combustion chamber at negligible velocity but leave it at 300 m/s. With this additional information does the answer to Question 4 require revision? If so, what is the new answer?

[1 503 kg/s]

Q.6. Calculate the stagnation enthalpy of air at 495 K if the enthalpy of the flowing air is 0·5 MJ/kg and the air is flowing with a velocity of

(a) 10 m/s,   (b) 100 m/s,   (c) 1000 m/s

[(a) 500·05 kJ/kg; (b) 505 kJ/kg; (c) 1000 kJ/kg]

Q.7. A 3 kW kettle holds 0·002 m³ of water. If the kettle is filled with water at a temperature of 15°C and 5 per cent of the total energy added is lost from the kettle by convection and radiation, how long does the kettle take to start boiling? Consider the specific internal energy per degree of the water to be 4 kJ/kg K.

[238 s]

Q.8. Consider a single stage of an axial flow impulse gas turbine. The specification for the turbine is that the axial velocity of the gas as it leaves the fixed blades which deflect the gas through 65° is 280 m/s, and the speed of the moving blades on the rotor is 260 m/s. If the moving blades deflect the gas through 100°, calculate the angle relative at the turbine axis and velocity at which the gas leaves the moving blades relative to the stationary parts of the turbine. (Figure 9.8 (c) may help with the velocity diagram construction.)

[14·5°, 289 m/s]

10

Derived equations

*The steady-flow energy equation appears in more than one form.
Euler presented it in a form useful in the study of flows of compressible fluids.
Bernoulli presented it in a form useful in the study of flows of incompressible
fluids. It is shown that these are both limited forms of the steady-flow energy
equation.*

10.1 Fluid flow

There are few branches of engineering in which fluid flow plays no part.
Gases including oxygen enter an internal combustion engine and leave it in the form
of products of combustion. Cooling water flows through the jackets of engines and
passes through the tubes of condensers and heat exchangers. There is the flow of
sea relative to a ship passing through or of air relative to an aircraft passing
through. Steam, water or gas flow through turbines that drive alternators in power
stations. Freon, ammonia or carbon dioxide flow around a refrigeration cycle.

The steady-flow energy equation (8.6) has a general application in all
these and in many other situations in engineering. Bernoulli's and Euler's equations, which are forms of that general equation, are often thought of as basic equations derived from fundamental principles. The fact that they are themselves
special forms of the steady-flow energy equation tends to be forgotten. It is re-
grettable from the engineer's point of view that these equations have been given
separate names, because this confuses the subject. Bernoulli's and Euler's forms
of the steady-flow energy equation are often presented as if some new principle
were involved, which makes the unwary think that in using one of these named equations he is doing something different and important.

10.2 The first law and a system

The first law is a law of conservation, expressed in equation (3.2) in
terms of energy and, ignoring for the present kinetic and gravitational energies,

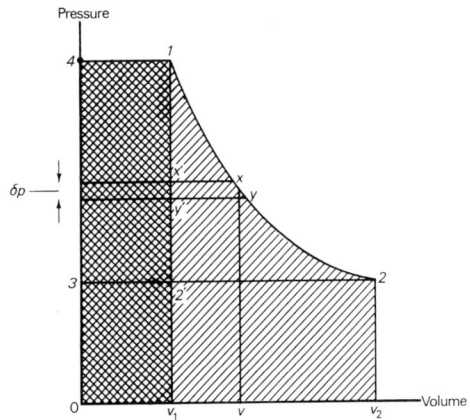

FIG 10. 1   p-v Changes for both compressible and incompressible fluids

can be written for an infinitely small change from, say, state x to y as in Fig. 10. 1:

$$\delta Q - \delta W = \delta U$$

or $$\delta Q = \delta U + p\delta V \qquad (10.1)$$

If one uses the applied mechanics definition of work

$$W = p\delta V$$

At this point we assume the energy changes of equation (10. 1) to have been carried out reversibly and so we write δQ_r instead of δQ. It is necessary to introduce a new property, **Entropy**, S, defined in Chapter 14, in terms of its change per unit mass of a system, as

$$T\delta s = \delta q_r$$

or in terms of the total entropy of the system

$$T\delta s = \delta Q_r \qquad (10.2)$$

Using equations (10. 1) and (10. 2) we may write down that for the reversible process between x and y

$$T\delta S = \delta U + p\delta V \qquad (10.3)$$

Equation (10. 3) is a statement of the changes in a system of properties which are dependent only on the end states (see section 2.3). Therefore equation (10.3) remains true whether the process is reversible or not. We will now consider the changes in the same system if it is in motion.

10.3 Systems in motion

The steady-flow energy equation, derived in Chapter 8, states that for a series of systems in motion,

$$Q - W_X = \Delta H + \Delta K + \Delta Z \qquad (8.6)$$

or, applied to a system passing through a control volume and changing from state x to y as in the previous section

$$\delta Q - \delta W_X = \delta H + \delta K + \delta Z \qquad (10.4)$$

It is useful to consider what happens to the enthalpy H when the state of the fluid changes from state x to state y. As can be seen from equation (8.7) the enthalpy change is

$$\delta H = \delta U + p\delta V + V\delta p \qquad (10.5)$$

and from this and equation (10.4)

$$\delta Q - \delta W_X = \delta U + p\delta V + V\delta p + \delta K + \delta Z$$

and from this and equation (10.2)

$$T\delta S - \delta W_X = \delta U + p\delta V + V\delta p + \delta K + \delta Z$$

or using equation (10.3)

$$-\delta W_X = V\delta p + \delta k + \delta Z$$

If the system under consideration has a mass of one unit

$$-\delta w_X = v\delta p + \delta k + \delta z \qquad (10.6)$$

10.4 Euler's and Bernoulli's equations

The applications of both Euler's and Bernoulli's equations are limited to conditions in which w_X by work is negligible. Therefore equation (10.6) is written in both Euler's and Bernoulli's conditions of application as

$$0 = v\delta p + \delta k + \delta z \qquad (10.7)$$

or $$-v\delta p = \delta\left(\frac{V^2}{2}\right) + \bar{z}g \qquad (10.8)$$

where V signifies velocity of flow. Equation (10.8) is called Euler's equation (in fluid mechanics this equation is derived from force-momentum relations and conservation of mass).

If this small elementary change from x to y is integrated over the larger process from 1 to 2 in Fig. 10.1 equation (10.8) becomes

$$-\int_1^2 vdp = \frac{V_2^2 - V_1^2}{2} + (\bar{z}_2 - \bar{z}_1)g \tag{10.9}$$

It is of interest to see, for both compressible and incompressible fluids, the curve $1 - 2$ and the integral $\int_1^2 vdp$ on a p-v graph, and these are shown in Fig. 10.1. For fluids v is equal to $1/\rho$ where ρ is the density of the fluid, and equation (10.9) can then be written,

$$\frac{P_2}{\rho} + \frac{V_2^2}{2} + \bar{z}_2 g = \frac{P_1}{\rho} + \frac{V_1^2}{2} + \bar{z}_1 g \tag{10.10}$$

Equation (10.10) is called Bernoulli's equation.

10.5 The unsteady-flow energy equation

Equation (8.6), which we have referred to often in this chapter is written down in units of energy. Equation (8.14) is basically the same equation but in units of energy per unit mass per unit time. We were in Chapter 8 and earlier in this chapter dealing with the steady state. So we could therefore assume that the rate of flow of mass $\dot{M}_{Kero} + \dot{M}_{Air}$ mentioned as an example in equation (8.3) as entering the control volume of Fig. 8.2 at points 1 and 2 had the same total quantity as $\dot{M}_{Exh}$ leaving the control volume at point 3 in the same time. It would have been possible to write equation (8.14) in terms of energy per unit mass per unit time in this way

$$\dot{Q} - \dot{W}_X = \dot{M}_{Exh}(h_3 + k_3 + z_3) - \dot{M}_{Air}(h_1 + k_1 + z_1)$$
$$-M_{Kero}(h_2 + k_2 + z_2) \tag{10.11}$$

where h, k and z are respectively the enthalpy, kinetic and gravitational energies and the suffices 1, 2 and 3 refer to the states of the fluid at points 1, 2 and 3 of Fig. 8.2. If equation (10.11) is applied to the simpler case shown in Fig. 10.2, assuming for the moment that the *flow is steady*, equation (10.11) for this case would be written

$$\dot{Q} - \dot{W}_X = \dot{M}_2(h_2 + k_2 + z_2) - \dot{M}_1(h_1 + k_1 + z_1) \tag{10.12}$$

but, if the conditions of *flow were unsteady*, the only previously stated condition remaining being that nothing in the control volume except the working fluid could

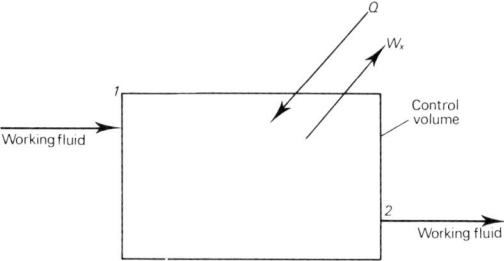

FIG 10.2   Unsteady flow through a control volume

store energy, equation (10.12) would have to be rewritten in the form,

$$\dot{Q} + \frac{d\dot{Q}}{dt}\,\delta t - \dot{W}_x - \frac{d\dot{W}_x}{dt}\,\delta t$$
$$= \left(\dot{M}_2 + \frac{d\dot{M}_2}{dt}\,\delta t\right)\left(h_2 + \frac{dh_2}{dt}\,\delta t + k_2,\text{ etc.}\right)$$
$$- \left(\dot{M}_1 + \frac{d\dot{M}_1}{dt}\,\delta t\right)\left(h_1 + \frac{dh_1}{dt}\,\delta t + k_1,\text{ etc.}\right) \qquad (10.13)$$

If some mass inside the control volume other than the working fluid could store energy and release it the equation would have to be further modified, but more information about such storage facilities would have to be known before any statement could be made about this.

10.6 Non-uniform flow

A further difficulty that arises in practice is that, due to friction, the fluid flowing into the control volume does not do so in a uniform way as we have assumed. Figure 10.3 shows a uniform pipe through which fluid flows. We are considering the flow at section **XX**. At any point x on the cross-section **XX** the distance from that point x on the cross-section to the point on the curve Z represents

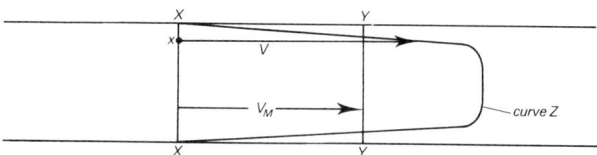

FIG 10.3   Changes of velocity across a cross-section

the velocity of flow at that point. We have said nothing about this variation of velocity of a fluid flowing through a pipe but, so far, have been working on the assumption that the velocity of flow V_M is uniform. This uniform state of affairs is shown in Fig. 10. 3 by the distance XY being constant, representing V_M. This has been of no consequence to our discussion but it is an incompleteness of which the reader should be aware. The true distribution of velocity V across the section XX varies in a way shown by the curve Z, starting at each wall at zero velocity because of friction and rising to a maximum in the middle. The exact shape of the curve depends on the type of fluid used and whether the flow is laminar or turbulent. The value of V_M in Fig. 10. 3 is the arithmetic mean value of V. The area between cross-section XX and the curve Z is equal to the area between cross-section XX and cross-section YY.

10. 7 Summary

The relationship of Euler's and Bernoulli's equations to the steady-flow energy equation was discussed and the unsteady-flow energy equation was also mentioned. For completeness the necessity of considering the velocity distribution of flows was discussed. There are no questions at the end of this chapter as the chapter merely serves as a link with the fuller study of fluid mechanics.

11

Processes in a heat engine

Now that we can consider steady-flow processes, the four processes that make up a heat engine are discussed. The similarities and the differences between the components of heat engines that use vapour such as steam for its working fluid are compared with those that use gas such as air for its working fluid. The components that make up a heat pump and a refrigerator are also considered.

11.1 The four parts of a heat engine

A heat engine is shown in Fig. 6.2 to be a system that takes in heat energy Q_1 at a high level of availability, and transforms part of it into work energy W and heat energy Q_2 at a lower level of availability. In Fig. 6.3 the processes that take place inside a heat engine are shown, whereby a working fluid flows around a cycle of processes and, in so doing, effects the transformation of energy. The four parts through which working fluid flows so undergoing four processes are:

(a) Heat exchanger 1 which takes in energy Q_1 from a high-temperature reservoir and transfers it to the working fluid.

(b) Adiabatic expander where the working fluid gives up work energy W_{xt}, or in the case of a heat pump is throttled in which case W_{xt} is zero.

(c) Heat exchanger 2 which takes the energy Q_2 from the working fluid and transfers it out of the heat engine to a low-temperature reservoir.

(d) Adiabatic compressor where the working fluid takes in energy W_{xp} and so returns the working fluid to its initial condition, thereby completing the cycle.

The process that occurs in each of these four parts is a steady-flow process in that in each case the mass flow-rate and stored energy of the working fluid at the same point within each part remains constant, not varying with time. The working fluid passes through each of the four processes in completing the overall cyclic process. Unsteady flow is not under consideration here, and therefore the processes may be examined by using the steady-flow energy equation.

11.2 A vapour power plant

In Fig. 11.1 the thermodynamic cycle of Fig. 6.3 is used for a practical purpose as a steam-power plant. One has to imagine the state of the working fluid being relatively cold at 1 where it is being forced into the boiler, which is heat exchanger 1. The boiler may be gas or oil-fired, receiving heat energy q_1 per unit mass of working fluid from the combustion products of the oil or gas, or it may be receiving q_1 from another fluid that has been through a nuclear reactor for example. The heat energy is transferred in the boiler to the working fluid changing

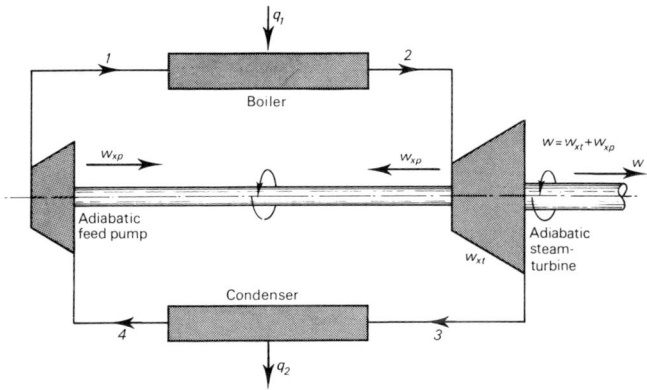

FIG 11.1   Steam-turbine power plant

the fluid from being a liquid at 1 to being a vapour at 2. From state 2 at point 2 the vapour enters the turbine where it does work energy w_{xt} on the shaft per unit mass of working fluid. The shaft has two functions. Firstly, it does work w_{xp} (negative) on the fluid in the adiabatic compressor or feed pump and, secondly, transmits the balance of work energy $(w_{xt} + w_{xp})$ out of the heat engine to do useful work. The fluid, having lost energy, leaves the turbine at 3 in the form of wet vapour (a mixture of water in liquid and vapour forms) at a fairly low temperature. In this state it enters heat exchanger 2, a condenser in a vapour power plant, and gives up an amount of energy q_2 to the cooling water in the condenser. The fluid, a liquid once more, leaves the condenser at 4 and enters the feed pump where it has work w_{xp} done on it and is restored to its original state of cold water at boiler pressure.

11.3 A gas power plant

In Fig. 11.2 the thermodynamic cycle of Fig. 6.3 is used in such a way that the working fluid is always gaseous and the cycle is that of a closed gas-power plant. (The reason for 'closed' will become clearer in a later chapter.) The state of the working fluid at 1 is that of a relatively cold gas which is being forced into the heater or combustion chamber which in this case is heat exchanger **1**.

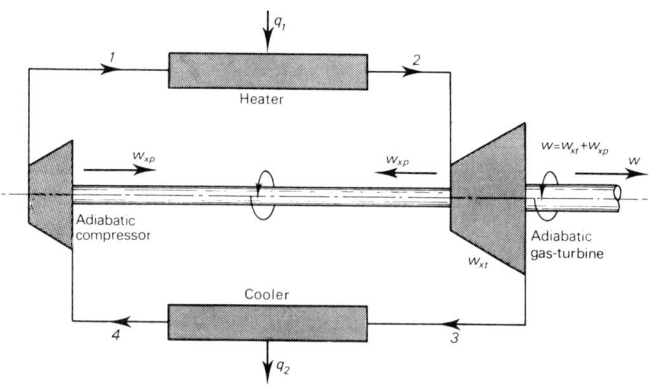

FIG 11.2   Gas-turbine power plant

The heater is externally fired by gas or oil, the products of combusion passing over the outside of the tubes through which the working fluid passes. If the heater itself is the combustion chamber into which fuel is sprayed so that it combusts with the oxygen of the working fluid the plant ceases to be strictly a system, which a heat engine should be, in that a small amount of mass is crossing the boundary. For the present it is sufficient for us to note this deviation and to assume that the effect of the additional mass is negligible—however the resultant energy release is by no means negligible. Whichever way it is done, the result is that, within heat exchanger **1**, the working fluid changes from a relatively cold to a hot gas at 2. From state 2 at point 2 the gas enters a turbine where it does work w_{xt} on the shaft. The gas, having lost energy w_{xt}, leaves the turbine at 3 as a relatively cold gas at low pressure and temperature. In this state it enters heat exchanger 2 and there gives up energy q_2 to the cooling atmosphere or other cooling fluid. The working fluid leaves the heat exchanger **2** at 4 and enters the compressor where work w_{xp} is done on it, energy being thereby added to the gas in restoring it to its initial condition at 1. So the cycle is complete.

11.4 A refrigerator or heat pump

In Fig. 11.3 the reversed thermodynamic cycle of Fig. 6.5 is used for a practical purpose as a refrigerator or a heat pump. One has to imagine the state

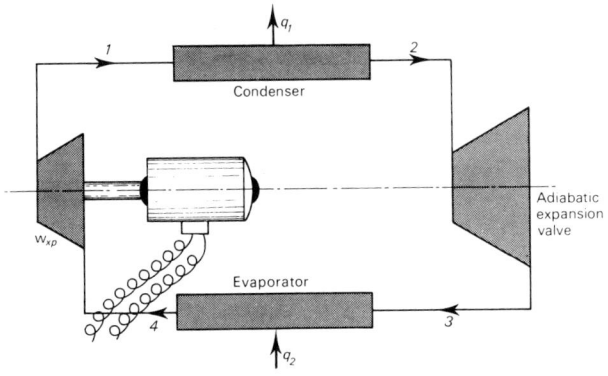

FIG 11.3   Refrigerator or heat-pump plant

of the working fluid as being that of a vapour at 1 where it is entering heat exchanger 1 which is a condenser in this case. The condenser may be cooled by water or by air, sending out heat energy q_1 per unit mass of working fluid to the cooling water or air, and leaving the condenser at 2 in a liquid state. From 2 the liquid enters an expansion nozzle where it expands at constant enthalpy (see section 9.2) to a lower pressure and leaves the nozzle at 3 as a vapour at the lower pressure but with the same enthalpy that it had at 2. A consequence of this throttling is that the temperature at 3 will be lower than the temperature at 2. The fluid, from 3, then enters heat exchanger 2 which is an evaporator in which it takes heat energy from a cold reservoir—from the cold chamber if the plant is being used as a refrigerator, or from a low-temperature reservoir if the plant is being used as a heat pump. The working fluid, having gained energy in the evaporator, leaves it at 4 without increase of pressure and then enters the adiabatic compressor in which its state changes until it returns to its original state at 1. No work is done in the adiabatic expansion valve and so the net external work w is w_{xp} which, because it is work done on the system by its surroundings, will numerically be negative.

Power and refrigeration plants (Q and A)

Q.1. What are two fundamental differences between steam-turbine and refrigerator plants?

A.1. (a) W for a turbine plant is positive but for a refrigerator plant is negative.

(b) The expander in a power plant is a unit doing positive work like a turbine, but in a refrigerator it is an expansion valve in which no work is done.

Q.2. Why in a steam-turbine and a gas-turbine plant does all the energy added as Q_1 not appear as W_{xt}?

A.2. See Planck's version of the second law in section 7.3 or consider what happens to the level of availability of the energy.

Q. 3.   When the first law is applied to the thermodynamic cycles of Figs. 11.1, 11.2 and 11.3, what is the common result?

A. 3.   All are cyclic plants and for each

$$\Sigma Q - \Sigma W = 0$$

Q. 4.   A study has been made in this chapter of ideal plants. If the plants were practical instead of ideal, list the conditions existing in the ideal plant that would then not hold—giving reasons.

A. 4.   (a) The turbine and compressor would not be completely adiabatic in that energy would be lost by heat to the surroundings because of the temperature difference between the system and the surroundings.

(b) Similarly the pipe work in the system would have energy losses by heat.

(c) There would be frictional losses in the system—effectively causing throttling of the working fluid because the insides of the pipes, turbines, etc., are not ideally smooth.

11.5 Summary

The conception of a magic box exchanging heat energy with its surroundings and doing work energy has been transformed into a working fluid undergoing four processes which if considered overall amounts to a system undergoing a cyclic process. The cycle is described in which a fluid is used to perform the thermodynamic cycle of a vapour power plant, gas to perform the cycle of a gas power plant, and fluid to perform that of a refrigerator are described.

11.6 Questions for the reader

Q. 1.   If the energy supply to the boiler of the steam-turbine plant of Fig. 11.1 is 450 kJ/kg of working fluid and the useful work w is 190 kJ/kg, what is the plant's efficiency?

[0·42 from equation (6.8)]

Q. 2.   In the plant mentioned in Question 1 what is the value of q_2?

[260 kJ/kg]

Q. 3.   If 1 kg/s of cooling water enters the jackets of a condenser at 10°C and leaves at 70°C, what is the value of Q_2 in that plant? c_p for water is 4·2 kJ/kg K.

[252 kJ/s or kW]

Q. 4.   If the efficiency of the plant of Question 3 is 0·33, what is its power?

[124 kW]

Q. 5.   A domestic refrigerator is pressed into service to make fifteen iced-lollies. Assuming that each iced-lolly is made of 0·1 kg of water and that the water is at

0°C when it is put into the refrigerator—if the coefficient of performance of the refrigerator is 4·0 and a 0·25 kW motor is used to compress the working fluid—calculate how long the children must wait for their frozen lollies. The energy given up when water freezes is 2 100 kJ/kg.

[3 150 s]

Q. 6   A man who has a large refrigerated cellar in a centrally heated house has the idea of constructing a heat pump between the cellar and his bedroom. If the cellar is kept at −13°C and his living room at +17°C assuming that he has a reversible heat pump what would be its coefficient of performance? (Equation 12. 2 may help).

[9·7]

Q. 7.   If in Question 6, 2 kW of power is removed from the cellar what would be the power rating of the compressor?

[0·23 kW]

Q. 8.   Do any of the following plants, when considered alone, fall within the definition of heat engines in the strict sense in which we have defined the term heat engine?

(a)   An aircraft jet engine
(b)   A steam plant in a generating station
(c)   An internal combustion engine in an automobile
(d)   A domestic refrigerator
(e)   An air conditioning plant

[(b) Yes, (d) and (e) Yes but reversed, (a) and (c) No—changes of mass]

12

The Carnot cycle

The theoretical heat engine of Carnot is defined. It is shown that the most efficient engine which can operate between two energy reservoirs at constan&#xfffd; temperature is that in which the working fluid undergoes a Carnot cycle. From th&#xfffd; and the idea of a reversible engine the idea of an absolute temperature scale is introduced.

12.1 Carnot's thermodynamic cycle

Carnot specified a thermodynamic cycle that consisted of a series of four non-cyclic processes which could be undergone by the working fluid in a reversible engine—that is to say in the best engine (see section 7.7) that could operate between two reservoirs at given constant temperatures. All processes in the series were reversible processes, together forming the reversible cycle show&#xfffd; in Figs. 12.1, 12.2 and 12.3. The processes were these:

    (a) From $1 \rightarrow 2$, a reversible isothermal process. This is a simple heat-transfer process in which the working fluid remains at a constant temperature throughout the process. No work is done. In this particular process the temperature remains at T_1, the temperature of the high-temperature reservoir, and heat energy q_{1r} is transferred to the working fluid.

    (b) From $2 \rightarrow 3$, a reversible, adiabatic process. This is a simple work-transfer process in which the temperature of the working fluid falls while work energy w_{xt} is transferred out of it at the expense of its stored enthalpy. No heating is done. It is due to the fluid's loss of enthalpy that the temperature of the working fluid falls.

    (c) From $3 \rightarrow 4$ a reversible isothermal process during which heat energy q_{2r} is transferred out of the fluid at constant temperature T_2&#xfffd;

(d)   From 4 → 1 a reversible adiabatic process during which work energy w_{xp} is transferred to the fluid.

All these processes are shown in Figs. 12.1, 12.2 and 12.3. Figure 12.1 is a temperature-entropy graph of which one axis is temperature T on a yet to be determined absolute scale and the other is the axis of another property called entropy s. This property, of which more will be written in succeeding chapters, is

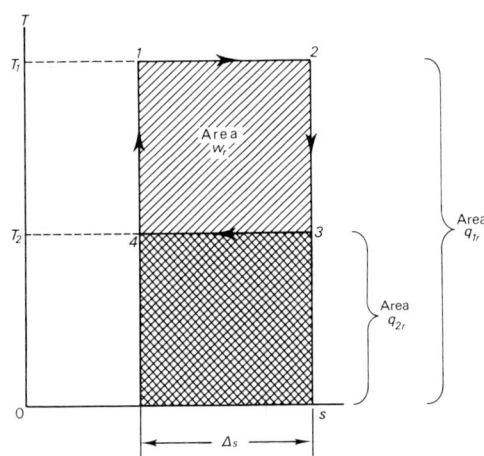

FIG 12.1   Diagram, showing Carnot's cycle with vertical axis representing temperature T and a horizontal axis such that the area under any curve representing a reversible state path is proportional to the energy transferred by heat during process

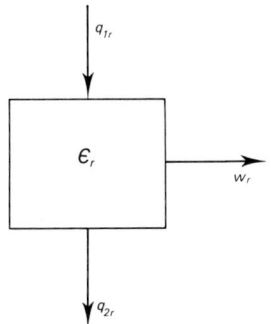

FIG 12.2   A reversible engine

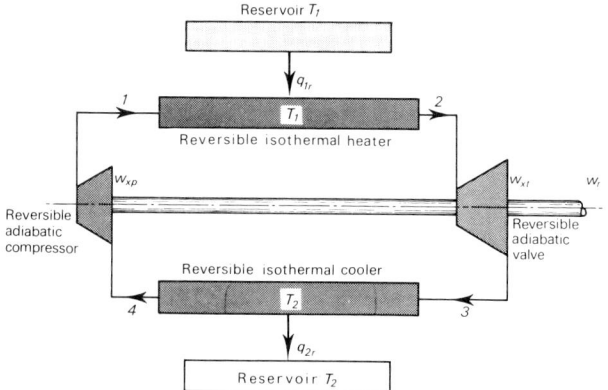

FIG 12.3   Processes in Carnot's engine

such that if a reversible state path is drawn on a temperature-entropy graph the
area under the curve will be proportional to the heat energy transferred during the
reversible change of state (see Chapter 14). The area under line $1 - 2$ in Fig. 12.1
is equal to q_{1r} and is numerically positive (heat energy is defined as +ve if it
enters the system), whereas the area under line $3 - 4$ is proportional to q_{2r} and is
numerically negative. The work w_r, being equal to $q_{1r} + q_{2r}$, as stated in equation
(6.3), is proportional to the area bounded by $1 - 2 - 3 - 4 - 1$ in Fig. 12.1.
Because $2 - 3$ and $4 - 1$ are adiabatic (i.e. have no heat transfer) as well as reversible the areas under these lines are zero. The statement that the area $T_1 \Delta s$
under the curve representing a reversible state path is equal to the heat energy
q_{1r} transferred reversibly leads one to the conclusion that, for process $1 - 2$

$$q_{1r} = T_1 \Delta s$$

see Fig. 12.1, and

$$q_{2r} = T_2 \Delta s$$

From these two equations and equation (6.8),

$$\eta_r = \frac{w_r}{q_{1r}}$$

from equation (6.3)

$$= \frac{q_{1r} + q_{2r}}{q_{1r}} \qquad\qquad 12.1)$$

$$= \frac{T_1 \Delta s - T_2 \Delta s}{T_1 \Delta s}$$

Dividing top and bottom by Δs, we get finally

$$\eta_r = \frac{T_1 - T_2}{T_1} \qquad (12.2)$$

It follows that in a cycle as shown in Fig. 12.1 that

$$\Delta s = \frac{q_{1r}}{T_1} = \frac{q_{2r}}{T_2} \qquad (12.3)$$

The negative sign of q_{2r} has been dropped because in equation (12.3) we are dealing with transfer by heat of energies q_{1r} and q_{2r} reversibly at T_1 and T_2. These transfers are reversible because they are between systems at the same temperature as will be explained in section 13.3; T_1 in the case of q_{1r} and T_2 in the case of q_{2r}. Equation (12.3) is therefore true no matter in which direction the energy is being transferred.

It is shown in section 7.7 that a reversible engine has the greater efficiency when compared with any irreversible engine working between the same two thermal reservoirs at constant temperatures T_1 and T_2. Hence it follows that equation (12.2) gives the greatest efficiency of any engine working between the two reservoirs T_1 and T_2.

12.2 Absolute scales of temperature

Equation (12.2) shows the greatest possible efficiency that can be exhibited by an engine working between two reservoirs at temperatures T_1 and T_2. It can be seen from this equation that the efficiency approaches 100 per cent—which according to the first law cannot be exceeded—as T_2 approaches zero. This does not mean zero on an arbitrary scale, such as the scale usually called Celsius, but zero on some absolute scale. This is because arbitrary scales such as Celsius and Fahrenheit would give different values of the efficiency for a given engine which is an absurd situation. For instance, if T_1 were 100°C or 212°F and if T_2 were 20°C or 68°F, the efficiency from equation (12.2) would be $(100 - 20)/100 = 0.80$ on the Celsius scale and $(212 - 68)/212 = 0.68$ on the Fahrenheit scale. Our concept of efficiency is not something that should depend on our temperature scale, so how do we need to define our temperature so that we get unique values for efficiency? Both Celsius and Fahrenheit scales have been arbitrarily devised; zero on the Celsius scale being chosen for the melting point of saturated ice in air at a pressure of one atmosphere (a little more than 10^5 N/m^2 or 1 bar), whereas zero on the Fahrenheit scale was chosen as the temperature at which a saturated solution of ice and common salt melts.

From such ideas arose the concept of a scale of temperature starting from some absolute zero. The problem being to define the absolute zero point.

12.3 Absolute zero

To find real or absolute zero one must imagine a series of reservoirs (see Fig. 12.4) having constant temperatures T_1, T_2, T_3, etc., down through any arbitrary zeros to an absolute zero, assuming such a point exists. Between each reservoir and the next below it in temperature there is a reversible heat engine operating. The spacing between the temperatures of the reservoirs is such that the work energy W given out by each engine is the same. Consider the first engine

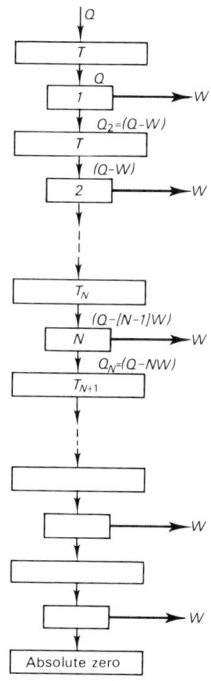

FIG 12.4   A series of energy reservoirs and reversible engines.

in the series. From the first law, because heat energy transferred in is Q and the work energy transferred out is W it follows that the heat energy transferred to the second reservoir, at T_2, is $(Q - W)$ as shown in Fig. 12.4. Because it is an engine we have applied the first law assuming there is no change in the stored energy of the engine, $\Delta U = 0$, or in other words the process undergone by the working fluid is cyclic.

If the work energy transferred out of the second engine is W it follows that the heat energy transferred to the third reservoir, at T_3, is $(Q - 2W)$. The work done by each engine is the same and one can say about the Nth engine, between reservoirs at temperatures T_N and T_{N+1}, that the heat energy transferred to it from the reservoir at T_N is $\{Q - (N - 1)W\}$, the work transferred out of the engine is W, and the heat energy transferred out to the reservoir at temperature T_{N+1} is $(Q - NW)$.

From equation (12. 3) we know for the first engine that

$$\frac{Q}{T_1} = \frac{Q - W}{T_2} \tag{12. 4}$$

that for the second engine

$$\frac{Q - W}{T_2} = \frac{Q - 2W}{T_3} \tag{12. 5}$$

and that, for the Nth engine

$$\frac{\{Q - (N - 1)W\}}{T_N} = \frac{(Q - NW)}{T_{N+1}} \tag{12. 6}$$

From these last three equations we may write

$$\frac{Q}{T_1} = \frac{Q - W}{T_2} = \ldots = \frac{Q - (N - 1)W}{T_N} \tag{12. 7}$$

It can also be shown from equation (12. 2) that

$$\eta_1 = \frac{T_1 - T_2}{T_1}; \eta_2 = \frac{T_2 - T_3}{T_2}; \eta_N = \frac{T_N - T_{N+1}}{T_N} \tag{12. 8}$$

From equation (6. 1), and because W is the work energy transfer from each engine,

$$W = \eta_1 Q = \eta_2 (Q - W) = \ldots = \eta_N\{Q - (N - 1)W\} \tag{12. 9}$$

From equations (12. 8) and (12. 9) eliminate η_1, η_2, etc., and hence obtain

$$W = (T_1 - T_2) \frac{Q}{T_1} = (T_2 - T_3) \frac{Q - W}{T_2} = \ldots$$

$$= (T_N - T_{N+1}) \frac{Q - (N - 1)W}{T_N}$$

From this last equation and equation (12. 7) we get

$$(T_1 - T_2) = (T_2 - T_3) = \ldots = (T_N - T_{N+1}) \tag{12. 10}$$

Equation (12.10) tells us that the temperatures of the reservoirs, which we specified as being such that reversible engines between adjacent pairs of reservoirs would give out equal work W, are all equally spaced. The fact that the reservoirs are equally spaced in temperature makes their temperatures very suitable for use as a scale of temperature. The temperature scale so defined is called the ABSOLUTE TEMPERATURE SCALE. It goes down to absolute zero as will be explained in this section.

It can be observed in Fig. 12.4 that the heat energy transferred to engine 1 is Q and the heat energy leaving engine 1 is $(Q - W)$ and leaving engine 2 is $(Q - 2W)$. Similarly the energy leaving engine 3 is $(Q - 3W)$ and engine N is $(Q - NW)$. In fact as each engine is passed the energy being transferred decreases by an amount W at every step. Finally, we find the residual energy leaving the penultimate engine of the series is W and that leaving the last engine is zero. In order that equation (12.7) may be applied to the last engine the temperature of the last reservoir must be zero—absolute zero. This is consistent with what we think of a reservoir at absolute zero temperature. While it is at absolute zero it has no energy in store and no energy is transferred to it. An absolute thermodynamic scale may be defined as follows.

Let the number N of engines or spaces between reservoirs be the number of degrees of temperature in the scale from T_1 in Fig. 12.4 down to absolute zero then the temperature differences are

$$(T_1 - T_2) = \ldots = (T_N - T_{N+1}) = 1 \text{ degree}$$

and the number N of degrees down to absolute zero will be

$$N = \frac{Q}{W} \tag{12.11}$$

The practical implication of an absolute temperature scale is that one point, absolute zero, is fixed implicitly. To define a temperature difference on an absolute scale one arbitrarily chosen additional fixed point is required. The point chosen is the point at which water exists as a solid, liquid and vapour, the triple point (see Chapter 17). It is chosen so that the degree is about the same size as the degree Celsius and the triple point is found to be at 273·16 degrees Kelvin (K) above absolute zero. The scale is given Lord Kelvin's name because it was he who first used the series of engines with equal work output described earlier in this section to derive an absolute scale of temperature. A similar absolute scale can be derived in which the degree is about the same size as the degree Fahrenheit and the triple point is then found to be 460 degrees Fahrenheit above absolute zero (sometimes called degrees Rankine).

The absolute scale (Q and A)

Q.1. Assume T_1 the temperature of the first in a series of reservoirs such as the series shown in Fig. 12.4, has a known value and that the heat energy transferred into and out of that reservoir is 200 kJ. Assume also that the work energy

W from each engine in the series is 2 kJ. If the last reservoir has a temperature of absolute zero, what is the temperature of the next to last reservoir?

A.1.   The number N of degrees of temperature from T_1 down to absolute zero is given by equation (12.11) as

$$N = \frac{Q}{W} = \frac{200}{2} = 100$$

The intervals of temperature between T_1 and zero must be $T_1/100$. If the last reservoir has a temperature of absolute zero, the next to last must have a temperature of $T_1/100$ degrees.

Q.2.   If T_1, the temperature of the first reservoir, is 100°C and the temperature intervals between reservoirs are to be 1°C downward to absolute zero, how many engines would there be in the series?

A.2.   On the absolute Celsius, or Kelvin scale the temperature of the reservoir, T_1, would be 273 + 100, so if the temperature interval is 1°C per engine, then there will be 373 engines.

Q.3.   If in the last question the work energy rate transferred by each engine is 0·268 kJ/s, what are $\dot{Q}_1$ and $\dot{Q}_2$ for the last engine; what is this engine's efficiency and how many degrees below the temperature of the first reservoir—the one at 100°C—is absolute zero?

A.3.   From Fig. 12.4, Q_2 is zero because no energy can be transferred to a 'reservoir' that can store no energy—in this case a reservoir at absolute zero, and therefore because $\dot{W}$ is 0·268 kJ/s, $\dot{Q}_1$ is 0·268 kJ/s, and the efficiency of the engine is 100 per cent. On an absolute scale the freezing point of water, about 0·02°C different from the triple point, can be taken to be 273 degrees above absolute zero, and therefore the numbers of degrees that the reservoir at absolute zero is below one at 100°C is 373 degrees usually called degrees Kelvin.

(*Note:* It is implied in Answer 3 that when one reservoir is at absolute zero of temperature it is possible to obtain 100 per cent efficiency. However, as soon as any energy is transferred to a reservoir at absolute zero its temperature must rise. This paradox which in reality is not of importance to engineers is sometimes stated in the third law of thermodynamics—the fact that it is impossible to obtain absolute zero temperature.)

12.4 Summary

It has been stated that Carnot's cycle of thermodynamic processes undergone by a working fluid in a reversible engine consists of two reversible isothermal and two reversible adiabatic processes. The efficiency of such an engine was shown to be given by

$$\eta_r = \frac{T_1 - T_2}{T_1}$$

From this equation a relationship between the two heat transfers associated with a reversible engine was shown to be

$$\frac{q_1}{T_1} = \frac{q_2}{T_2}$$

A series of engines with Carnot efficiency was used to establish an absolute scale of temperature going down to absolute zero.

12.5 Questions for the reader

Q.1. How many fixed points, not counting absolute zero, are required to define (a an arbitrary, and (b) an absolute temperature scale?

[Two, one]

Q.2. A new arbitrary temperature scale is defined with its zero at the point whe solid carbon dioxide sublimes, −78°C, and at the point where sulphur boils, 444°C. If on this new scale there are one thousand degrees between the fixed points, find the temperature on the new scale corresponding to 0°C.

[149]

Q.3. If on a new temperature scale, 100 corresponds to 27°C and 200 correspon to 327°C is this new scale an absolute scale or an arbitrary scale?

[Absolute]

Q.4. Knowing pv = RT is the law obeyed by a perfect gas, what would a p − v diagram look like when a perfect gas is the working fluid undergoing a Carnot cycle?

[See Fig.12.5]

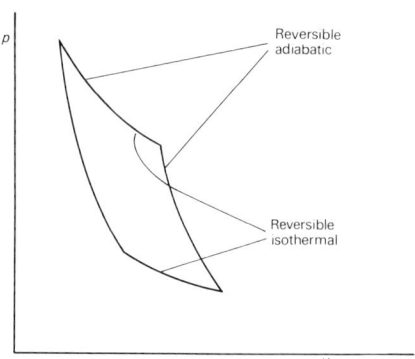

FIG 12.5   Answer to question 4

Q. 5. In Fig. 12. 6 below there are two systems of engines operating between two reservoirs at temperatures of 227°C and 27°C. System (a) involves two Carnot engines using an intermediate reservoir at 127°C, and system (b) involves one Carnot engine. What are the efficiencies of the two engines involved in (a), and the efficiency of (b); overall which is the best system (a) or (b)?

$$[\eta_1 = 0\cdot2; \eta_2 = 0\cdot25; \eta_3 = 0\cdot4; \text{overall the same}]$$

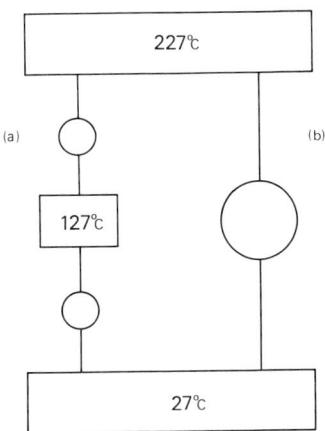

FIG 12. 6   To illustrate question 5

Q. 6. In Question 5 what would be the number of engines in systems (a) and (b) between the upper reservoir at 227°C and absolute zero if the work energy transferred from each was equal?

[Five for (a) but (b) does not give equal temperature spacing because the interval between 227°C and 27°C is 200°C and the interval between 27°C and absolute zero is 300°C. The work energy transfers from the engine of (b) could not therefore be equal and one cannot therefore answer the question with regard to (b)]

Q. 7. If in a series of reversible engines of equal work output the rate of energy transfer by heat to the first engine is 100 kW at 1000 K and there are 25 engines in the series, what is the rate of work output per engine and the temperature differences between the energy reservoirs?

[4 kW; 40 K]

Q. 8. In Kelvin's series of equal work output reversible engines what does a grap
of engine efficiency plotted against mean temperature of the reservoirs look like

[Consider eight engines between 800 K and absolute zero; see Fig.
12. 7]

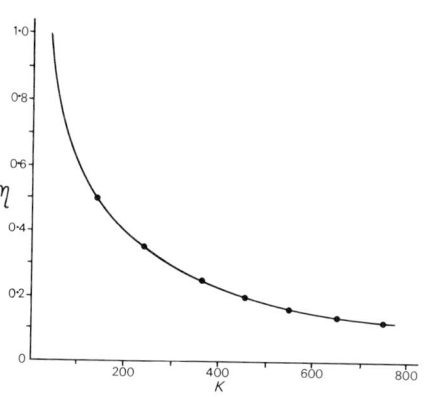

FIG 12. 7   Answer to question 8

13 The best processes

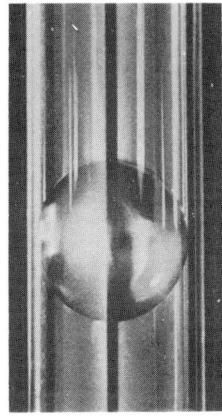

The earlier discussion in Chapter 7 of the reversibility of heat engines is now broadened to discuss the reversibility of processes and how this can be recognised. This is applied to simple processes involving only work or heat and to the general process where both are involved. Entropy is introduced.

13.1 The most efficient engine

In Chapter 7 the most efficient engine operating between two thermal reservoirs was proved to be a reversible engine. An engine was said to be reversible if, whilst taking in energy q_r supplied reversibly by heat, it gives out energy w_r reversibly by work, this being the same amount of work energy w_r as the engine would require to take in if it were reversed and used between the same two thermal reservoirs to give out the same energy q_r by heat.

A heat engine at its simplest may be thought of as a working fluid undergoing four processes linked together in such a way that the fluid goes through a cycle of processes. Four such processes forming a cycle are shown in Fig. 6.3 where they are seen to be

(a) between points 1 and 2
(b) between points 2 and 3
(c) between points 3 and 4
(d) between points 4 and 1

which are shown separately in Figs. 13.1 (a), (b), (c) and (d). Linked together they form a heat engine working on the thermodynamic cycle $1 - 2 - 3 - 4 - 1$ shown in Fig. 6.3. The best heat engine is a reversible engine. It is taken as axiomatic that a reversible engine must comprise only reversible processes. What is a reversible process?

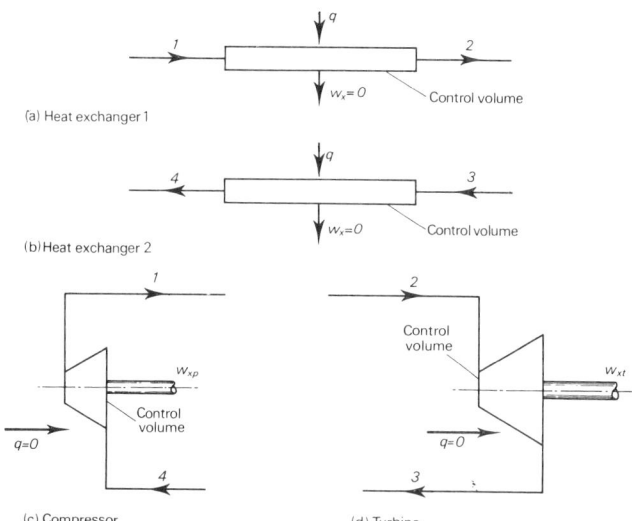

FIG 13.1   Processes that are parts of a heat engine cycle

13.2 The general process

The processes (a), (b), (c) and (d) mentioned in section 13.1 and show in Fig. 13.1 are each special cases of a general steady-flow process like that show in Fig. 13.2 (a) which has three features. As the fluid passes through the control volume there occur,

(a) A transfer to the fluid of heat energy q from surroundings at temperature T
(b) A transfer from the fluid of work energy w_x to the surroundings
(c) A change of stored energy h, k and z in the fluid equal to any inequality between q and w.

Equation (8.13), the steady-flow energy equation is applicable

$$q - w_x = \Delta h + \Delta k + \Delta z \qquad (8.1?)$$

but to simplify the discussion without in any way altering the principle we are assuming that the changes Δk and Δz in kinetic and gravitational energies are negligible and that the steady-flow energy equation can therefore be written, for the process in Fig. 13.2 (a), as

$$q - w_x = \Delta h$$
$$= h_b - h_a \qquad (13.$$

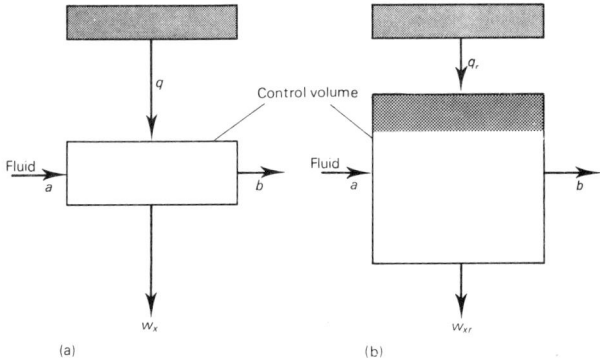

FIG 13.2   Two general steady-flow processes

The above brief description of a general process does not answer the question 'What is a reversible process? Let us now consider how equation (13.1), which describes the general process would change if the process were a general reversible process.

Clearly, for comparison's sake, the end states of the fluid must remain the same if we are going to compare the general process which may be irreversible with the reversible general process. It is the heat and work actions that must change. In the special case of a reversible general process the heat energy, q_r not q, must be transferred reversibly and the work energy, w_{xr} not w_x, must also be transferred reversibly. Our question 'What is a reversible process?' can only be answered for any particular change of state $(h_b - h_a)$ by asking the question 'What are reversible heat and work actions?'

13.3 A reversible heat transfer action

Reversibility in heat transfer is fairly easily identified with the help of the second law, which states that energy is transferred by heat from a system at one temperature to a system at a lower temperature. This law is thought to be true for transfer in one direction equally as for transfer in the other. It follows that, for a heat action to converge on reversibility, the action must converge on one in which the heat energy is transferred along zero temperature gradient between the surroundings and the system. This is the only type of heat transfer that can operate equally well in either direction.

This situation is impossible to achieve because it would require an infinitely large heat exchanger. However, it is the situation towards which improvers of heat exchangers must strive if they seek the best heat exchanger. That is why in Fig. 13.2 (b) the reservoir from which q_r is taken is shown with the same intensity of shading as that of its point of contact with the control volume, signifying equal temperatures, and it is also why the control volume of Fig. 13.2 (b) is much larger than that of Fig. 13. 2 (a) signifying that the reversible heat exchanger is infinitely large.

To sum up, a reversible heat transfer occurs between a system and its surroundings when energy is transferred along a zero temperature gradient. As this is an impossibility all heat exchangers can only approximate to this ideal.

13. 4 A reversible work transfer action

Work energy w_x in Fig. 13. 2 (a) is being transferred in such a way that it could conceivably be stored outside the control volume wholly by the raising of a mass in a gravitational field (see section 3. 3). In the case of w_x in Fig. 13. 2 (a) and also in the case of w_{xr} of Fig. 13. 2 (b), these energies once outside the control volume could each be stored wholly by the raising of a mass. Because these energies have been transferred by work all the energy could be returned to the system at its original level of availability. So how does one determine which i a reversible work-transfer action? The answer is that it cannot be determined in the obvious way that is possible with a heat-transfer action.

The only way in which one can determine whether or not a given work-transfer action is reversible is to reverse the process and look at the result. Later when the concept of entropy is introduced it will be seen that changes of entropy may be used to judge the reversibility of a work-transfer action.

This may not seem to be an entirely satisfactory picture but perhaps it is clarified if an example is used to illustrate the point.

Consider a turbine, such as that shown in Fig. 13. 1 (d), and imagine that the working fluid passing from 2 to 3 produces work energy w_{xt}. Now let us assume that there is no energy transfer by heat, which would be possible if the thermal lagging were 100 per cent effective. So the process occurring between 2 and 3 can either be reversible or it can be irreversible depending solely on whether the work is reversible or irreversible. There is no heat. Assume that we can isolate the problem of the reversibility of work to depend on the behaviour of the bearing supporting the shaft. If the process is reversible the behaviour of the bearing must be reversible; that is to say it must be frictionless. The bearing would not behave reversibily if there were friction because the friction would cause a temperature difference due to energy dissipation. This would mean that some of the energy being transferred in the bearing would not be transmitted as work but dissipated as an increase in internal energy. In the real situation it will be irreversible. If the working fluid changes irreversibly between states 2 and 3 the work energy produced will be less than for the reversible case because the bearing, on account of its inherent friction, will prevent the maximum work transfer. To justify the assumption about the bearing would be an involved process but it is sufficient to say that all losses are irreversible.

13. 5 The reversible general process

Consider a fluid that undergoes a process in steady flow, the most important feature of which is a change of state in that its enthalpy changes from h_a t

h_b. In this respect—in respect of the fluid's end states—the process is identical with the general process discussed in section 13.2 and is therefore comparable with that process. In a real situation the process would be carried out irreversibly as shown in Fig. 13.2 (a) with both heat and work actions irreversible, but the ideal would be to do it reversibly as in Fig. 13.2 (b). Let us consider the reversible case. As the fluid passes through the control volume there occur,

> (a)   A reversible transfer to the fluid of heat energy q_r from surroundings at temperature T. Reversibility is achieved by the temperature T of the surroundings always being the same as the temperature of the fluid with which it is in thermal contact. Note that this can only be achieved by using a heat exchanger of infinite size (see section 13.3).
>
> (b)   A reversible transfer from the fluid of work energy w_{xr} to the surroundings. Reversibility is achieved by storing all the work energy w_{xr} when it has gone into the surroundings by the raising of a mass (see section 13.4).
>
> (c)   A change of stored energy h, k and z in the fluid equal to any inequality between q_r and w_{xr}.

The process we are now discussing has been carried out reversibly and therefore the steady-flow energy equation is applicable in the form,

$$q_r - w_{xr} = h_b - h_a \qquad (13.2)$$

if we continue to assume that Δk and Δz are negligible. Equation (13.2) for the reversible case causing a change $(h_b - h_a)$ of stored energy should be compared with equation (13.1) for the irreversible case causing the same change $(h_b - h_a)$ of stored energy. The common term in equations (13.1) and (13.2) is $(h_b - h_a)$ which describes the change of state we are talking about.

13.6 Some reversible processes

There are three reversible processes worth considering in detail. These are shown on a temperature-entropy graph in Fig. 13.3 by curves 1, 2 and 3. The corresponding flow diagrams are shown in Figs. 13.4 (a), (b), and (c). Entropy is a property of a fluid that will be discussed fully later. Each of the processes represented by curves 1, 2 and 3 are reversible and in each case the fluid enters the control volume in state a and leaves in state b so that the overall change $(h_b - h_a)$ of enthalpy is the same in each case.

In process no. 1 of Figs. 13.3 and 13.4 (a) the fluid in state a enters a heater in which it receives energy q_{1r} isothermally and reversibly. Isothermal heating of a gas is difficult to achieve but isothermal heating of a wet vapour is not difficult. If, for instance, the heater were a boiler within which the water changed from a liquid to a vapour at constant atmospheric pressure the change would take

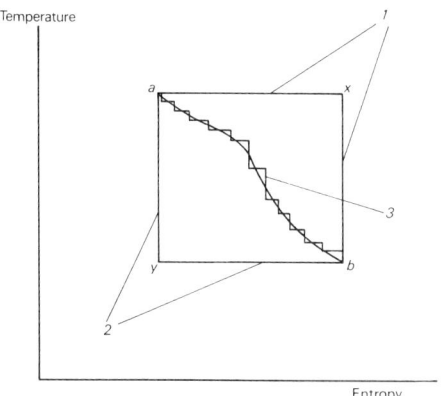

FIG 13. 3   Three reversible processes

place at 100°C. It is impossible to achieve reversible heating of either a gas or a vapour because this could only be done in an infinitely large heat exchanger. In process no. 1 the fluid would leave the heat exchanger in state x and enter a reversible turbine in which its state would change from x to b while it gave out work energy w_{xr}.

In process no. 2 of Figs. 13. 3 and 13. 4 (b) the fluid in state a enters a reversible adiabatic turbine in which its state changes from a to y. The fluid in state y then enters a reversible isothermal heater. If the heater were a boiler for instance in which the fluid changed from liquid to vapour at constant pressure the change would be isothermal. The same difficulty in achieving isothermal heating of a gas exists in process no. 2 as it did in heating in process no. 1. The fluid in state b leaves the cooler and the control volume at point b.

Process no. 3 of Figs. 13. 3 and 13. 4 (c) is represented by a curve. The processes of heating or cooling and working are not carried out in one stage each, as they were in processes nos. 1 and 2. If there were a large number of heaters, coolers and compressors arranged in an appropriate order a curve of any degree of smoothness and of any shape beginning at a and ending at b could be drawn. The first reservoir for the first heat exchanger would be at T_a and the last would be at T_b. By selecting the number of heaters and of coolers and compressors and turbines one could arrange that the curve no. 3 was of any desired shape.

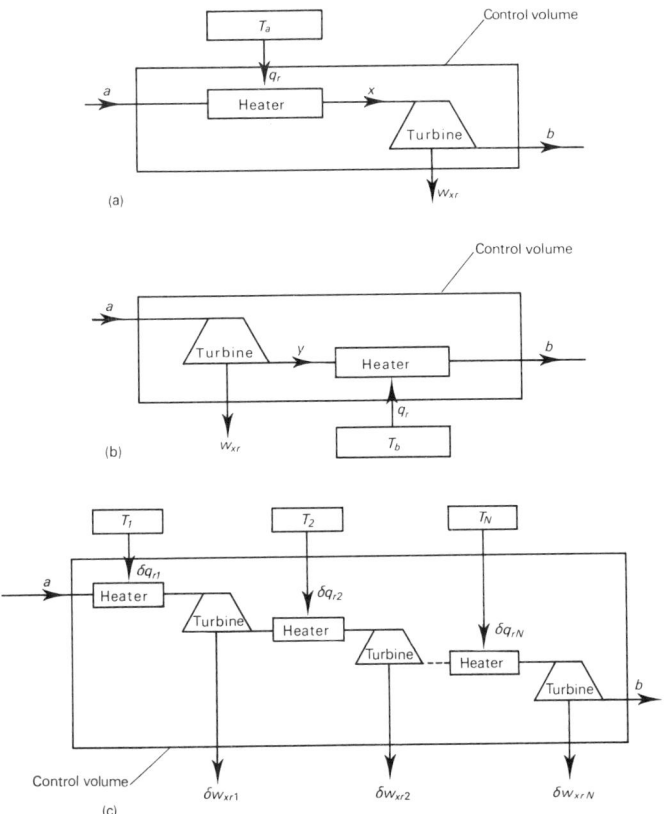

FIG 13.4   (a) Reversible process no. 1. (b) reversible process
no. 2. (c) reversible process no. 3.

13.7 Summary

In a study of processes a general process has been described with its
two actions, heat and work, which transfer stored energy into and out of fluid passing through a control volume. The identification of reversible heat and work and
finally a reversible general process with its two actions, reversible heat and reversible work, that transfer stored energy into and out of fluid passing through a
control volume, were discussed.

13.8 Questions for the reader

Q.1. In a refrigerator that is reversible what are the four processes that combin to make up the cycle?

[(1) Evaporator,   (2) Compressor,   (3) Condenser,   (4) Turbine or reversible nozzle]

Q.2. Give two reasons why in practice a reversible heat transfer is impossible.

[(1) An infinitely large heat exchanger cannot be obtained and (2) Energy cannot be transferred along zero temperature gradient by heat]

Q.3. In reversible heat and work transfers what happens to the overall level of availability of the energy?

[By definition it must remain constant]

Q.4. A chestnut is roasted in front of the fire and while the chestnut is at a high temperature the fire goes out. The chestnut once cold became hot and cold again. Is this a reversible heat transfer?

[No, because (a) The energy is transferred along a temperature difference and (b) There is an internal energy change within the chestnu

Q.5. On a cloudless night a pool of water is subject to a temperature of −0·01°C and by morning some ice has formed on the pool—is this a reversible heat transfe

[Very nearly!]

Q.6. A massive sail is raised by means of a series of pulleys and cables, and the work energy done is stored as gravitational energy in the sail. What property mus this system have for the process of raising the sail to have been reversible?

[The pulleys must be frictionless]

Q.7. Can you think of any practical reversible process that involves heat and wor transfers?

[No, not truly reversible]

14 *Entropy*

Entropy is defined and shown to be a property. The entropy changes for simple and general reversible and irreversible processes are considered. Entropy changes for processes on the Universal Scale are briefly considered.

14.1 Entropy

Consider a fluid undergoing a general process in the course of which the state of the fluid changes from state a to state b while in thermal contact with an energy reservoir at a particular temperature. This change can be carried out reversibly as shown in Fig. 13.2 (b) or irreversibly as shown in Fig. 13.2 (a). In both cases the change is the same, from a to b, as enthalpy is a property depending only on the end states of the process. h_a the enthalpy in state a, and h_b the enthalpy in state b, are the same for the reversible and irreversible cases. The changes of state from a to b are shown in Fig. 14.1 where the axes of the graph represent:

    (a)   Absolute temperature T
    (b)   Another parameter s

The nature of the parameter s is defined in such a way that the area (shown cross-hatched) under the line representing a reversible state path (shown as a full line) is equal to the heat energy transferred reversibly between the control volume and its surroundings. The parameter s is called **Entropy,** and is known, for reasons given later in this chapter, to be a property of the fluid. A change or absence of change in this property can be used to determine whether or not an adiabatic process has been carried out reversibly.

14.2 Entropy and any process

It should be firmly in the reader's mind that, irrespective of what actually happens, for which of course only one true path can be drawn, the change

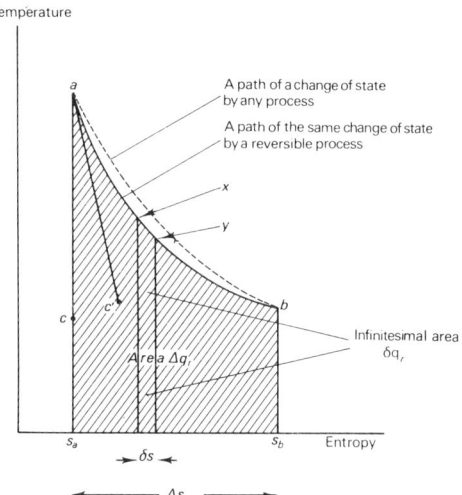

FIG 14.1   The path of a change of state and a reversible path of the same change of state

from a to b could have been along any one of an infinite number of reversible and irreversible paths that could have been drawn on the T-s graph. The link connecting all reversible paths on the T-s diagram is the area beneath the curve representing the reversible path. Consider the small change of state from x to y in Fig. 14.1. The area under this element xy of the curve must be equal to the heat energy that would have been transferred if the change had taken place reversibly.

     The link connecting all paths whether reversible or not is their having the same end states, and therefore the same overall Δs for the change. This is because entropy is a property and, as discussed in section 2.3, the change of a property depends only on the end states of the system, not at all on the process joining these end states.

14.3 Definition of entropy

     In Fig. 14.1, consider a small change of state from state x to state y on a reversible path—one of the infinite number of small changes of state in terms of T and s that together make up the whole change from a to b if the change had taken place along that reversible path. The area δq_r, under that small part of the curve, is an infinitesimal part of the total transfer of heat energy Δq_r that would be transferred in the course of a reversible change from a to b. In terms of T and s one can write

$$\delta q_r = \text{area under curve x} - \text{y}$$
$$= T\delta s$$

or $\qquad \delta s = \dfrac{\delta q_r}{T}$ $\hspace{3cm}$ (14.1)

If the change in s during the change from a to b is Δs one can write:

$$\Delta s = s_b - s_a$$
$$= \int_a^b ds$$

or $\qquad = \int_a^b \dfrac{dq_r}{T}$ $\hspace{3cm}$ (14.2)

from equation (14.1).

Entropy is in fact defined by equation (14.2) in terms of its change Δs.

14.4 The best adiabatic process

From the point of view of a process forming part of the best engine which operates in a thermodynamic cycle the best adiabatic (simple work transfer) expansions and compressions are those that produce the maximum work energy for the cycle. We know that for this purpose the engine must be a reversible engine, and hence the processes must also be reversible. This means that the expansions and compressions must not only be adiabatic they must also be reversible and so occur with no change of entropy. Because it is reversible for the whole process, the energy change by heat during every small part of the process is δq_r, and, because it is adiabatic δq_r equals zero. Therefore,

$$\delta q = \delta q_r = 0$$

and from equation (14.2)

$$\Delta s = s_b - s_a$$
$$= 0$$ $\hspace{3cm}$ (14.3)

This means that there is no change of entropy during the best—the reversible—adiabatic process. An adiabatic and reversible process is therefore **ISENTROPIC** and it follows that of all adiabatic processes a reversible one can be recognised by being isentropic. This is illustrated on Fig. 14.1 by the line ac which is a reversible process, whereas the process ac' is not isentropic as $s_{c'} \neq s_a$.

In this section it has been stated that the best adiabatic process (that is to say the most suitable as a component of the reversible engine) is a reversible adiabatic process and in that case the process is isentropic for which, from equation (14. 3),

$$\Delta s = 0$$

14. 5 Entropy and an irreversible adiabatic process

The fact that $\Delta q = 0$ in an adiabatic process that has been carried out irreversibly is not inconsistent with Δq_r, for the same changes of stored energies having a finite value. Δq is the heat energy that was actually transferred while the stored energies in the system were changed from h_a, k_a and z_a to h_b, k_b and z_b, whereas Δq_r is the heat energy that would have been transferred if the process were reversible but resulted in the same changes of stored energy as the irreversible process. The inference can be drawn from this is that if a process is adiabatic and is irreversible the same process carried out reversibly would not be adiabatic.

In an irreversible adiabatic process carried out between any two states, $q = 0$ but $q_r \neq 0$, and so for such a process, from equation (14. 2),

$$\Delta s \neq 0 \qquad\qquad\qquad\qquad (14.$$

In an adiabatic process, therefore either equation (14. 3) or (14. 4) is true—this in itself is seen not to be very restricting! In addition to this we can state that for adiabatic irreversible processes of the kind in which we are interested the following can be written

$$\Delta s > 0 \qquad\qquad\qquad\qquad (14.$$

but the following can never be written

$$\Delta s < 0$$

The general statement that is being made about adiabatic process in a heat engine is

$$\Delta s \geqslant 0 \qquad\qquad\qquad\qquad (14.$$

which is a combination of equations (14. 3) and (14. 5).

Before proving equation (14. 6) we will prove that entropy is a property.

14. 6 Proof that entropy is a property

Suppose there are three reversible processes that all cause the same change of state from a to b (see Fig. 14. 2). Equation (14. 2) tells us that, for the

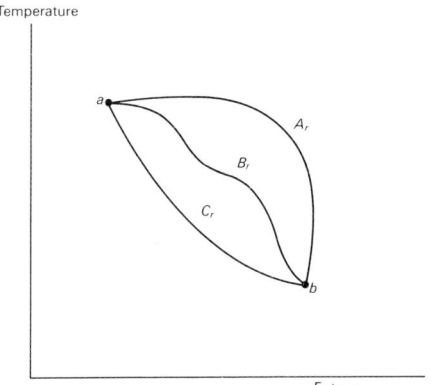

FIG 14. 2   Reversible changes of state

cycle a A_r b B_r a,

$$\Delta s = \int_{A}^{b} \frac{dq_r}{T} + \int_{B}^{a} \frac{dq_r}{T}$$

and because we are discussing a cycle the property s will have the same value after the cycle is complete and so

$$\int_{A}^{b} \frac{dq_r}{T} + \int_{B}^{a} \frac{dq_r}{T} = 0$$

similarly, for the cycle a A_r b C_r a,

$$\int_{A}^{b} \frac{dq_r}{T} + \int_{C}^{a} \frac{dq_r}{T} = 0$$

If we subtract one of these equations from the other we get

$$\int_{B}^{a} \frac{dq_r}{T} - \int_{C}^{a} \frac{dq_r}{T} = 0$$

that is,

$$\int_{B}^{a} \frac{dq_r}{T} = \int_{C}^{b} \frac{dq_r}{T}$$

So that the value of $\int_{a}^{b} \frac{dq_r}{T}$ — that is to say from equation (14. 2) the value of the change of entropy — is the same whichever reversible path is taken from a to b. If entropy s depends on the end states only, not on the path along which the changes take place, then entropy is a property, and $\Delta s = (s_b - s_a)$ has the same value for any path reversible or not.

14.7 A cyclic process

A cyclic process is a process or a series of processes undergone by a fluid whereby the state of the fluid at the end of the process is the same as it was at the beginning. A common cyclic process is that undergone by the working fluid a heat engine in the series of processes shown in Figs. 6.3 and 14.3. The processes forming the series are:

> 1 − 2 A simple heat transfer process,
> 2 − 3 An adiabatic process,
> 3 − 4 A simple heat transfer process, and
> 4 − 1 An adiabatic process.

These together form a cyclic process undergone by the fluid, which is initially in state 1, takes in heat energy q_1, gives out work energy w_{xt}, also gives out heat energy q_2, and finally takes in work energy w_{xp} which restores its state again to state 1. This is a thermodynamic cycle, a cyclic process, and takes place in what is called a heat engine.

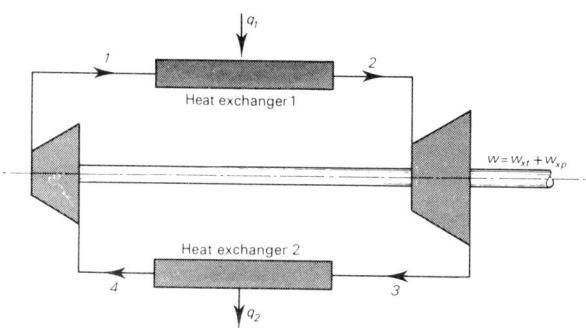

FIG 14.3   A cyclic process

The first law (section 2.5) states that for one cycle, in terms of units of energy per unit mass of working fluid, that

$$\oint dq_1 + \oint dq_2 - [\oint dw_{xt} + \oint dw_{xp}] = 0 \tag{14.7}$$

or $q_1 + q_2 - w = 0$

if q_1 and q_2 are the heat energies and w the total work energy $(w_{xt} + w_{xp})$ transferred between the fluid and its surroundings. For such an engine it was shown in section 6.4 that its performance is given by its efficiency η where

$$\eta = \frac{w}{q_1}$$

$$= \frac{q_1 + q_2}{q_1} \tag{14.8}$$

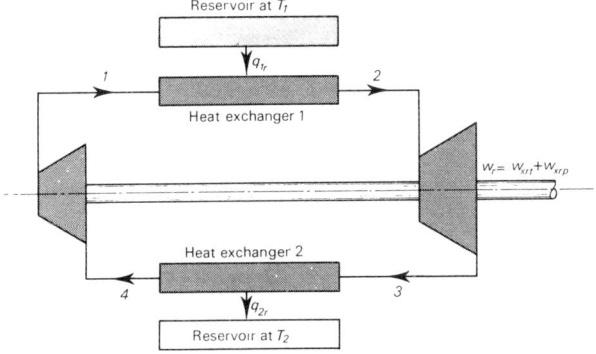

FIG 14.4   A reversible cyclic process

Similarly for a reversible engine (see Fig. 13.2) working between reservoirs at the same temperatures as those mentioned above and taking in reversibly the same quantity of heat energy, $q_{1r} = q_1$, then for the reversible engine equation (14.7) becomes

$$\oint dq_{1r} + \oint dq_{2r} - [\oint dw_{xt} + \oint dw_{xp}] = 0 \tag{14.9}$$

or $q_{1r} + q_{2r} - w_r = 0$

and the equation of efficiency becomes

$$\eta_r = \frac{w_r}{q_{1r}}$$

$$= \frac{q_{1r} + q_{2r}}{q_{1r}} \tag{14.10}$$

We have shown in Chapter 7 that an irreversible heat engine is less efficient than a reversible one, or

$$\eta < \eta_r$$

and therefore, for the thermodynamic cycles of the engines in Figs. 13.1 and 13.2, equations (14.8) and (14.10) give

$$\frac{q_1 + q_2}{q_1} < \frac{q_{1r} + q_{2r}}{q_{1r}} \qquad (14.11)$$

When comparing the two engines' performance it is useful to compare them for the same q_1. Therefore

$$q_1 = q_{1r} \qquad (14.12)$$

or $\qquad \dfrac{q_1}{T_1} = \dfrac{q_{1r}}{T_1} \qquad (14.13)$

and from the two equations (14.11) and (14.12)

$$q_2 < q_{2r}$$

or $\qquad \dfrac{q_2}{T_2} < \dfrac{q_{2r}}{T_2} \qquad (14.14)$

From equations (14.13) and (14.14) it follows that

$$\frac{q_1}{T_1} + \frac{q_2}{T_2} < \frac{q_{1r}}{T_1} + \frac{q_{2r}}{T_2}$$

which can be rewritten in the cyclic form

$$\oint \frac{dq}{T} < \oint \frac{dq_r}{T} \qquad (14.15)$$

Because the term $\oint \dfrac{dq_r}{T}$ is the change of entropy from the definition by equation (14.2), and because the term is integrated over the whole cycle the term must be zero. That is,

$$\oint \frac{dq_r}{T} = 0 \qquad (14.16)$$

and, from (14.15) and (14.16),

$$\oint \frac{dq}{T} < 0 \qquad (14.17)$$

It follows from equations (14.16) and (14.17) that if q is a variable standing for the energy transferred either reversibly or irreversibly at any point on a cyclic process that,

$$\oint \frac{dq}{T} < 0 \qquad (14.18)$$

This is known as Clausius' inequality.

14.8 Units of entropy

The units of entropy may be easily deduced from equation (14.2) which states

$$\Delta s = \int \frac{dq_r}{T}$$

Because dq_r is in units of energy per unit mass and T in units of temperature, s must be $\dfrac{energy}{mass \times temperature}$ or more specifically kJ/kg K

14.9 Entropy and the universe

Consider how when a process occurs the entropy of the Universe changes. If we consider the Universe to be infinite then one must consider any processes that occur within the Universe to be adiabatic if the Universe is considered as the system.

Then for any change within the Universe the entropy change Δs is always greater than zero because any practical process is irreversible. The consequence of this fact is that energy becomes less available as processes occur, or to put it another way as time passes.

14.10 Summary

The best engine is a reversible engine and, to be reversible, the engine must have reversible processes. Two actions—heat and work—occur during a process. The recognition of reversibility in heat and work actions has been discussed. The idea of entropy was introduced and its relationship to reversible and irreversible processes described. Because entropy is a property its change during a process is the same whether or not the process takes a reversible path. Entropy has been defined and related to simple heat and simple work-transfer processes. In particular it has been stated that during a simple work-transfer, or adiabatic process, entropy remains constant or increases. In a cyclic process

$$\oint \frac{dq}{T} < 0$$

14.11 Questions for the reader

Q.1. How would you judge whether a simple work-transfer process was ever reversible?

[By examining what happens to the entropy of the fluid undergoing the process]

Q. 2. Consider a 3 kW kettle. The water in the kettle is just liquid at 100°C and is vaporised and it takes 10 min for it all to boil, what is the entropy change of the water during the boiling process?

[4·83 kJ/K]

Q. 3. If 10 kg of water takes 125 min to boil in the kettle in Question 2 what is the entropy change in kJ/kg K?

[6·03 kJ/kg K]

Q. 4. Water at an initial enthalpy of 210 kJ/kg passes through a generalised process where it emerges with an enthalpy of 2 600 kJ/kg. If kinetic and gravitational energies are negligible, and if the water is flowing at 0·2 kg/s with an energy transfer of 1 MW to the water what is the work transfer during the process?

[+522 kw]

Q.5. In what pieces of equipment do processes approximating to the following occur:

(a) A simple reversible work transfer
(b) A simple reversible heat transfer
(c) A reversible general process?

[(a) A turbine.   (b) A heat exchanger.   (c) An unconventional turbine with energy exchange]

Q. 6. If 5 kg of water at 20°C are mixed with 10 kg of water at 25°C what is the change of entropy for the mixture? (c_p = 4 kJ/kg K)

[0·00012 kJ/Kg]

Q. 7. What is the difficulty of working out an entropy change for Question 6?

[The process is a mixing process and hence irreversible]

Q. 8. If we consider the Universe in a classical sense (that described in section 14. 9) what is happening to the temperature of the Universe?

[(a) The overall mean temperature remains constant and (b) the differences in temperature become smaller]

Q. 9. What happens to the level of availability and the change of entropy during

(a) A reversible adiabatic process
(b) An irreversible adiabatic process?

[(a) Both remain constant.   (b) Level of availability decreases, change of entropy increases]

15 Entropy and a power plant

The property entropy has previously been shown to give an assess-ment of the reversibility of an adiabatic process. How it can be used quantitatively is described in this chapter when applied to the adiabatic processes occurring in a turbine and a compressor.

15.1 Power plants

In all power plants, for instance in the power plants shown in Figs. 14.3 and 14.4, the working fluid that goes around the thermodynamic cycle under-goes at least two adiabatic processes—those of expansion and compression. In an irreversible power plant such as that in Fig. 14.3 the adiabatic processes are not so suitable for the purpose as they are in the reversible power plant of Fig. 14.4. How suitable or unsuitable adiabatic processes are we assess by the change of entropy which occurs during the process. The best adiabatic processes, as was explained in section 14.4, are isentropic,

$$\Delta s = 0 \tag{14.3}$$

In the less than best adiabatic processes the change of entropy is greater than zero. The worse—less reversible—the process is the greater is the increase of entropy, never a decrease, because as stated in equation (14.6) for any adiabatic process

$$\Delta s \geqslant 0 \tag{14.6}$$

This is what we now intend to prove.

15.2 In any adiabatic process $\Delta s \geqslant 0$

An adiabatic process is one in which there is no heat and therefore, for such a process, during which a fluid in motion changes from state a to state b,

$$\int_a^b \frac{dq}{T} = 0 \qquad (15.1)$$

Suppose the state of a system is changed from a to b along an irreversible path D as shown in Fig. 15.1 and is then changed back from b by a reversible path A_r to a.

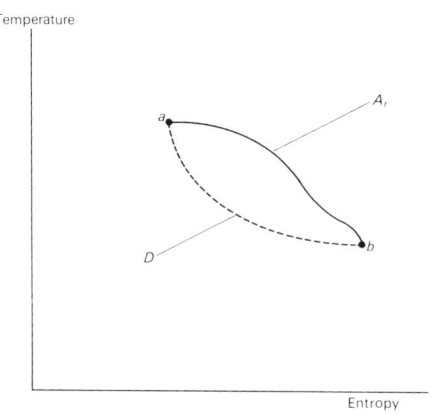

Temperature

Entropy

FIG 15.1   A cyclic process

As the cyclic change of state a—D—b—A_r—a is an irreversible cycle, because the path D is irreversible, we know from equations (14.16) and (14.17) that

$$\int_a^b \frac{dq}{T} + \int_b^a \frac{dq_r}{T} < 0$$

and, from this and equation (14.2), that

$$\int_a^b \frac{dq}{T} + \frac{\Delta s}{b \to a} < 0$$

or

$$\int_a^b \frac{dq}{T} - \frac{\Delta s}{a \to b} < 0$$

for the change of state from a to b, which for an irreversible adiabatic change of state from a to b becomes

$$0 - \frac{\Delta s}{a \to b} < 0$$

and so $$\frac{\Delta s}{a \to b} > 0 \tag{15.2}$$

Also from equation (15.1), which is for a reversible *adiabatic* change of state from a to be,

$$\left(\frac{\Delta s}{a \to b}\right)_r = 0 \tag{15.3}$$

These two equations (15.2) and (15.3) prove three statements that have been made earlier in this book. These are

(a) A reversible adiabatic process is isentropic—equation (15.3)
(b) During an irreversible adiabatic process the entropy increases—equation (15.2).
And, from (a) and (b) together,
(c) During an irreversible adiabatic change the entropy either remains the same or increases—equations (15.2) and (15.3) which together prove equation (14.6).

15.3 Turbines

In a gas power plant we usually think of a turbine being used as an expander, although reciprocating engines are also used as will be explained in a later chapter. The process taking place in the turbine is of a gas expanding from one pressure to another lower pressure. For instance, as shown on the temperature-entropy graph for a gas in Fig. 15.2, the gas might enter the turbine in state 2 indicated by point 2 on the graph and expand to state 3 at point 3 on the constant-pressure line 3′ − 3 − 5. Knowing that the process in the turbine is adiabatic we use equation (14.3) to state that the best—that is the reversible—state path would be isentropic and would therefore be represented in Fig. 15.2 by the state path 2 − 3. The work done by the fluid during the process would, from equation (13.2) (in which q_r would be zero because the process is adiabatic as well as reversible), be

$$w_{xrt} = -(h_3 - h_2) \tag{15.4}$$

If the process were not reversible, knowing that it is nevertheless adiatic we use equation (14.5) to state that the process would be represented in Fig. 15.2 by the

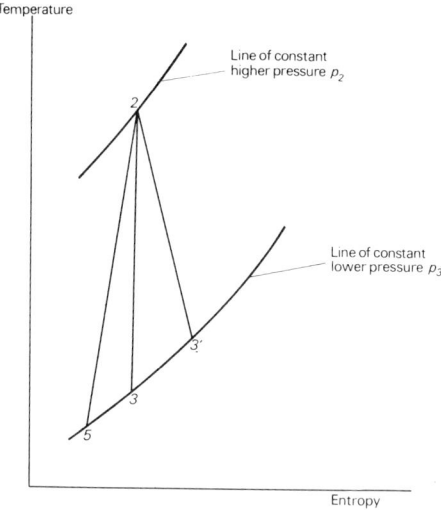

Temperature

Line of constant
higher pressure p_2

Line of constant
lower pressure p_3

Entropy

FIG 15.2   The paths of a change of state between two pressures

state path $2 - 3'$. The work done by the fluid during the process would, from equation (13.1) be

$$w_{xt} = - (h_{3'} - h_2) \qquad (15.5)$$

$w_{xt} < w_{xrt}$ because $h_{3'} > h_3$.

Equation (14.6) states with regard to adiabatic process that they can be as represented in Fig. 15.2 by lines similar to line $2 - 3'$ and, if the process is reversible too, by line $2 - 3$, but no adiabatic process can be represented by line $2 - 5$ because this displays a decrease in entropy, which cannot happen in an adiabatic process.

15.4 Compressors

In a gas power plant the process taking place in the compressor is of a gas being compressed from one pressure to another higher pressure. For instance, as shown on the temperature-entropy graph for a gas in Fig. 15.3, the gas might enter the compressor in state 4 indicated by point 4 on the graph and expand to state 1 on the constant pressure line $1 - 1'$. Knowing that the process in the compressor is adiabatic we use equation (14.3) to state that the best—that is the reversible—state path would be isentropic and would therefore be represented in Fig. 15.3 by the state path $4 - 1$. The work done on the fluid during the process would, from the steady-flow energy equation, equation (13.2), (in which q_r would

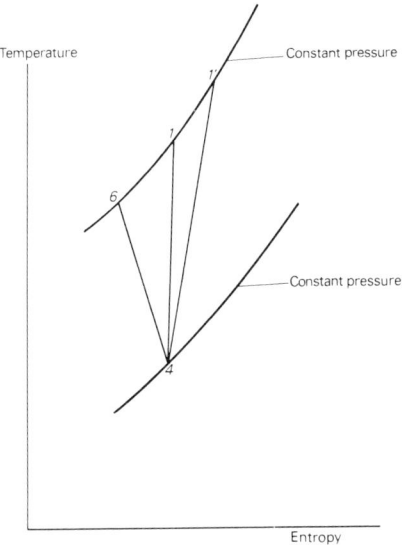

FIG 15.3   The paths of a change of state between two pressures

be zero because the process is adiabatic as well as reversible), be

$$w_{xrp} = - (h_1 - h_4) \tag{15.6}$$

If the process were not reversible, knowing that it is nevertheless adiabatic we would use equation (14.5) to state that the process would be represented in Fig. 15.3 by the state path $4 - 1'$. The work done on the fluid during the process would, from equation (13.1) (in which q would be zero because the process is adiabatic), be

$$w_{xp} = - (h_{1'} - h_4) \tag{15.7}$$

Equation (14.6) states with regard to adiabatic processes that they can be as represented in Fig. 15.3 by lines similar to line $4 - 1'$ and, if the process is reversible too, by lines $4 - 1$, but no adiabatic process can be represented by line $4 - 6$ because this displays a decrease in entropy, which cannot happen in an adiabatic process. In this case $w_{xp} > w_{xrp}$ because $h_{1'} > h_1$.

15.5   Isentropic efficiencies

The use of the word efficiency as we are now using it is not the same as the efficiency defined by equation (6.8) because it is not based on performance as defined by equation (6.7). The isentropic efficiency η_I of a turbine is a ratio given by

$$\eta_I = \frac{\text{work actually done by a turbine}}{\text{work that would be done if turbine were isentropic}}$$

$$= \frac{w_{xt}}{w_{xrt}} \tag{15.8}$$

where w_{xt} and w_{xrt} are given by equations (15.4) and (15.5).

The isentropic efficiency η_I of a compressor is a ratio given by

$$\eta_I = \frac{\text{work that would be done on a compressor if it were isentropic}}{\text{work actually done on a compressor}}$$

$$= \frac{w_{xrp}}{w_{xp}} \tag{15.9}$$

It should be noted that both these quantities, equations (15.8) and (15.9), are less than unity and (15.8) uses the isentropic work as the denominator whereas (15.9) uses it as the numerator. The use of these two isentropic efficiencies is shown in the following questions and answers.

Isentropic efficiencies (Q and A)

Q.1. The enthalpy per unit mass of a fluid entering a turbine is 400 kJ/kg and it would be 189 kJ/kg when leaving if the process inside the turbine had been isentropic. However, the turbine has an isentropic efficiency of 0·86. What is in fact the enthalpy of the fluid leaving the turbine?

A.1. For the isentropic process, from equation (15.4),

$$w_{xrt} = -(189 - 400)$$

$$= 211 \text{ kJ/kg}$$

From equation (15.8)

$$w_{xt} = \eta_I w_{xrt}$$

$$= 0·86 \times 211$$

$$= 181 \text{ kJ/kg}$$

From equation (15.5)

$$183 = -(h_{3'} - 400)$$

Enthalpy of the fluid leaving the turbine

$$h_{3'} = 400 - 183$$

$$= 219 \text{ kJ/kg}$$

Q. 2. The enthalpy per unit mass of a fluid entering a compressor is 140 kJ/kg. The enthalpy of the same fluid when leaving the compressor would have been 250 kJ/kg if the compressor had been isentropic but is in fact 280 kJ/kg. What is the isentropic efficiency of the compressor?

A. 2. From equation (15.6)

$$w_{xp} = -(280 - 140)$$
$$= -140 \text{ kJ/kg}$$

From equation (15.7)

$$w_{xrp} = -(250 - 140)$$
$$= -110 \text{ kJ/kg}$$

From equation (15.9)

$$\eta_I = \frac{w_{xrp}}{w_{xp}}$$
$$= \frac{-110}{-140}$$
$$= 0 \cdot 785$$

15.6 Summary

The adiabatic components of a power plant are mentioned with the changes of entropy that occur in them if they are reversible and if they are irreversible. It is then proved that during any irreversible adiabatic process the entropy increases and that any reversible adiabatic process is isentropic. Based on the steady-flow energy equation expressions are given that express the work done in a turbine in terms of the changes of enthalpy of the working fluid in both the reversible and irreversible cases. Expressions for similar work in compressors are given. The isentropic efficiencies of turbines and compressors are defined.

15.7 Questions for the reader

Q. 1. The states of a fluid are such that its enthalpy is 3 373 kJ/kg and its entropy 6·590 kJ/kg K on entering an adiabatic turbine and its enthalpy is 2 796 kJ/kg and entropy 6·352 kJ/kg K on leaving. Is this a consistent set of statements?

[No. As the process is adiabatic the entropy cannot decrease]

Q. 2. Reconsider Question 1 if the entropy on leaving were 6·590 kJ/kg K. Is this now a consistent set of statements?

[Yes, theoretically. This is the perfect reversible as well as adiabatic turbine all designers would like to find]

Q. 3. Reconsider the situation of Question 2 and state what work the turbine would do, in kJ/kg.

[577 kJ/kg]

Q. 4. The enthalpy and entropy of fluid entering the same turbine are 3 373 kJ/kg and 6·590 kJ/kg K and respectively 2 900 and 6·100 on leaving it. What work does the turbine do?

[473 kJ/kg]

Q. 5. What is the isentropic efficiency of this turbine?

[0·823]

Q. 6. The states of a fluid are such that its enthalpy is 1 500 kJ·kg and its entropy 3·300 kJ/kg K on entering an adiabatic compressor and respectively 2 000 and 3·000 on leaving. Is this a consistent set of statements?

[No. See answer to Question 1]

Q. 7. Reconsider if the entropy on leaving were now 3·300 kJ/kg K. Is this now a consistent set of statements?

[Yes. See answer to Question 2]

Q. 8. Reconsider the situation of Question 6 and state what work would have to be done on the compressor, in kJ/kg.

[−500 kJ/kg]

Q. 9. The enthalpy and entropy of the fluid entering the same compressor are 1 500 kJ/kg and 3·300 kJ/kg K and respectively 2 200 and 3·460 on leaving it. What work is being done in the compressor?

[−700 kJ/kg]

Q. 10. What is the isentropic efficiency of the compressor of Question 9?

[0·715]

16

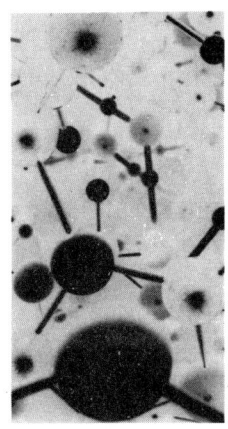

Liquids, vapours and gases

The working media undergoing processes in a heat engine are fluids.
It is therefore desirable that we should know what properties the fluids exhibit.
Fluid is the group term for three more specific classes, these being liquid, vapour
and gas. It is important to know into which class the fluid falls and the properties
of that class.

16.1 Matter

The layman is given to classifying matter in terms of what he thinks
he sees. He has little difficulty in deciding what is a solid, liquid or gas because
each has its own properties and each appears to the casual observer to behave con-
sistently. The scientist looks more deeply into matter and the more he looks, the
more his idea of separate distinct types of matter fades, so that the different types
of matter to use an analogy form a complete spectrum ranging from solid at one
end through liquids to vapours and gas at the other.

16.2 Solids and fluids

The engineer finds it difficult to distinguish between solids and fluids.
High velocity will make a relatively soft candle shot from a gun pass through a
solid wooden table. An aircraft of great mass can be supported safely by air if it is
moving sufficiently quickly. Copper and other metals pushed slowly through a hole
can appear to be flowing like a fluid, as also can the ice of glaciers and streams of
lava from a volcano. No satisfactory demarcation between solids and fluids can be
drawn without time playing an important part in the definition. One can say that
when one penetrates a substance one could do it slowly or quickly and the resistance of the substance to penetration could be very great or negligible. Matter is
called a **Fluid** when it appears to offer negligible resistance to slow penetration.
Matter is called a **Solid** when it appears to offer great resistance to slow penetration.

This demarcation between solids and fluids, given by these two definitions, serves very well because it takes account of the circumstances in which the
matter exists and allows matter to change with the circumstances from behaving
as a liquid to behaving as a solid—as it does when a stone at relatively high speed
is resisted by the water across the surface of which it is bouncing. The change of
behaviour from a solid to a liquid is exhibited by a metal as it flows through a die
like a liquid. Nevertheless the solidity of the initial and of the final object would
not be questioned.

16.3 Fluids

In section 16.2 we described the difference in behaviour of substance
which lead us to call some solids and some fluids. Although, in the study of thermodynamics, our primary interest is in fluids—that is to say with matter which offers
little resistance to slow penetration—it is appropriate here to mention in thermodynamic terms the relationship between matter behaving as a solid and matter
behaving as one of three types of fluid—liquid, vapour or gas.

Suppose one starts with a pure substance (not an alloy or solution)
strongly displaying the properties of a solid and puts energy into it. This could be
done by heat or by work, but let us suppose that in this case we add energy to the
system by heating it. The normal process of adding energy at a constant rate with
respect to time is shown as a temperature-time graph in Fig. 16.1. It must be

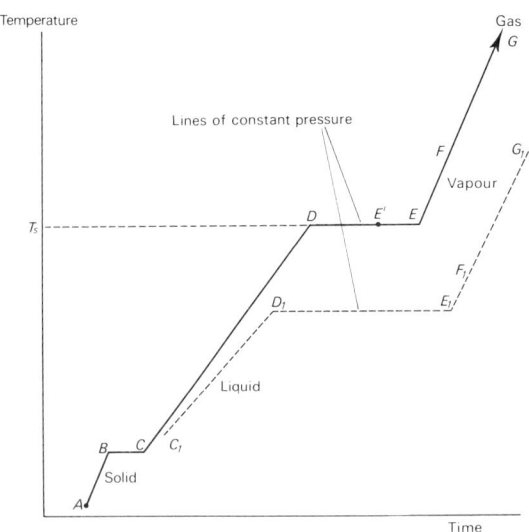

FIG 16.1   The state path of a fluid to which energy is being
transferred

imagined that at point A the system to which energy is to be added is solid. It is heated and its temperature rises with time until it reaches point B when the solid begins to melt—in other words it begins to change from a solid to a liquid. During the process of melting, which begins at B and ends at C, there is a relaxation in the bonding of the molecules which is bbserved to take place at nearly constant temperature as shown by the constant-temperature line BC on the graph. It is this relaxation in bonding that makes the system less resistant to penetration and more like a fluid than a solid. At C the change is complete and the system is now all liquid and no more energy can be stored at constant pressure without a rise of temperature. With further heating the temperature rises until the temperature at D is reached. Between C and D the fluid is a liquid. At D, with further heating, the fluid begins to change. B to C is the **first change of phase** and D to E is the **second change of phase.** At D no further energy can be stored without a second phase change; at D the fluid is said to be a **Saturated liquid.**

When more energy is added to the saturated liquid at constant pressure the liquid begins to evaporate which is the name given to the second change of phase during which the fluid changes from a saturated liquid at D to a vapour at E. During the process of evaporation there is again a change in the bonding of molecules which, if the pressure is kept constant, is also a constant temperature process as shown by the line DE on the graph. The constant temperature T_s of the fluid between D and E is known as the **Saturation temperature.** The energy that is taken into storage by the fluid without change of temperature is called **Latent internal energy** ΔU, or if changes of pV are taken into account **Latent enthalpy** ΔH. Sometimes it is called latent heat but this is a wrong use of the word heat because the transfer of energy to the fluid could be by heat or by work. At E the fluid is wholly vapour and one says that, having no liquid left, it is dry. Also it is saturated with energy in that no more energy can be stored in it at constant pressure without its temperature rising again, towards F. At E it is therefore called **Dry saturated vapour.** The fluid now shows properties not unlike those of a gas but we know that at E it is not a gas because it does not even approximately obey the ideal gas law—see section 18.2. A **Vapour** is a dry fluid that is at a temperature equal to or higher than the saturation temperature but which does not obey the ideal gas rule. At the state F in Fig. 16.1 the temperature of the vapour is higher than the saturation temperature and the vapour is said therefore to be **Superheated.** It is superheated in that its temperature is higher than the saturation temperature T_s. As more energy is put into store in the fluid with the pressure held constant the temperature rises still higher until at G, say, it sufficiently nearly obeys the ideal gas rule to be called a gas. It is not possible to specify a point on the line EG of Fig. 16.1 at which the fluid ceases to be a vapour and becomes a gas. All one can say is that, on a temperature-entropy graph, the further the state point is away from the critical point (see section 18.5) the more nearly does the fluid obey the ideal gas rule. When a fluid behaves sufficiently like an ideal gas for the discrepancies to be insignificant the fluid is called a gas and not a vapour.

It is important to remember that the process represented by the state path A B C D E F G has all taken place at constant pressure and the energy was added at a constant rate with respect to time.

Heating a fluid (Q and A)

Q. Steam is all liquid at state point D in Fig. 16.1 and all dry vapour at point E. Between these two it is referred to as wet vapour. If the changes of mass of liquid and vapour vary linearly from D to E of Fig. 16.1 with respect to time what is the wet steam's dryness at point E' where $DE' = 0 \cdot 8\ DE$?

A. The dryness at state point E' is proportional to DE'/DE because the change of mass varies linearly. Therefore the dryness, σ, is given by

$$\sigma = \frac{\text{mass of dry saturated vapour}}{\text{total mass of all the fluid present}}$$

$$= \frac{DE'}{DE} = \frac{0 \cdot 8\ DE}{DE} = 0 \cdot 8$$

16.4 Variations with pressure

In Fig. 16.1 it is seen that the fluid changes from D where it is a saturated liquid to E where it is a dry saturated vapour. If the change from D to E takes place at constant pressure it also takes place at a constant temperature, T_s, called the saturation temperature (sometimes called the temperature of evaporation or the temperature of condensation). The temperature of condensation is the same as the temperature of evaporation if the process is carried out very slowly but in practise one gets effects such as supercooling. If this occurs dry saturated gas can exist at temperatures lower than the condensation temperature without condensation actually taking place. These are unstable situations from a thermodynamic point of view.

It has been stated in section 16.3 that the change of state shown by the state path A B C D E F G has all taken place at the same constant pressure. If the pressure in the system had been constant but lower, another state path A_1 B_1 C_1 D_1 E_1 F_1 G_1 would have been plotted. Drawn to scale A_1 and B_1 would very nearly have been the same as points A and B and so they have been omitted. The saturation temperature represented now by D_1 E_1 for the lower pressure is lower than that represented by D E for the higher pressure. In fact if one specifies the pressure one is specifying the saturation temperature as these are uniquely related for any one fluid. When evaporation occurs there is a loosening of bonds between molecules. This can be thought of as being caused by increased agitation of the molecules associated with increased temperature and so it is not surprising that a higher temperature and therefore greater agitation is required to break the bonds when the bonds between molecules are reinforced by a higher pressure on the surface of the fluid where evaporation is taking place.

If we were to draw lines C D E F G for a fluid, for different constant pressures, on a graph similar to that shown in Fig. 16.1, we would get the family of lines shown in Fig. 16.2. The line joining all the points D_1, D_2, etc., together is called a **LIQUID SATURATION LINE**, and the line joining all the points E_1, E_2, etc., is **THE VAPOUR SATURATION LINE**. As the pressure rises and so the saturation temperature rises these two lines come close together until they meet at a point

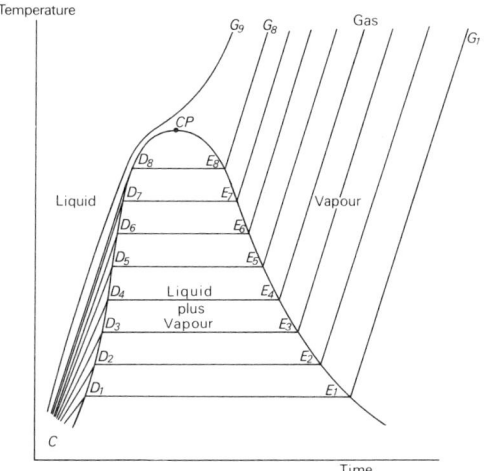

FIG 16. 2   Temperature—time curves for a fluid to which energy is
being transferred at different constant pressures

called the **Critical point**, CP. Classical experiments were carried out by Thomas
Andrews, 1813-85, on carbon dioxide in his investigations of the critical point.
        If the pressure on a liquid is such that the liquid will remain a liquid
at temperatures higher than the critical temperature then, at that pressure, addition
of more energy will cause the liquid to change to a dry vapour immediately without
the intervening situation when liquid and vapour coexist. Such a case is shown in
Fig. 16. 2 for the line ending at G_9. When at a certain pressure the temperature of
the gas has risen to a value higher than the saturation temperature T_S for that
pressure it is said to be superheated. When the state of the fluid in terms of pressure and temperature has risen above the critical point for that fluid the state of
the fluid is said to be **Supercritical.**
        We have in earlier chapters mentioned several properties in addition
to pressure and temperature. We have for instance mentioned specific internal
energy u (Chapter 3), specific enthalpy h (Chapter 8), and specific entropy s
(Chapter 14). The values of these properties are changing as the state point of the
fluid system moves along the path $C_1 D_1 E_1 F_1 G_1$ and they change by different
amounts along $C_2 D_2 E_2 F_2 G_2$.

16. 5 Property tables—water, liquid and vapour

        Property tables will be referred to frequently, and those in Ref. (3)
will be used in this book to find information required for the solution of many
problems. Property tables give values of properties not only for water but also for
other substances such as mercury, ammonia, freon 12 (dichlorodifluoromethane)

and other fluids of engineering interest. Steam and mercury are used as working fluids in power plants, and ammonia and freon 12 are used in refrigeration plants.

To keep our description brief we have used the suffixes 1 and v where referring to the properties of saturated liquid and dry saturated vapour respectively. Suffix lv refers to the change of a property between the two states. For instance h_1 is the specific enthalpy of the saturated liquid at point D in Fig. 16.1, h_v of dry saturated vapour at point E, and h_{lv} is the latent specific enthalpy for a change from D to E and is equal in value to $(h_v - h_1)$. In some tables f is used as a suffix for liquid and g for vapour.

Those parts of the tables showing the properties of steam usually have a column that gives the pressure at which all the other properties have the values shown in the same horizontal line, and another column that gives the saturation temperature T_s corresponding to that pressure. It may be seen in Table 16.1 that at a pressure of 1×10^5 N/m^2 which is approximately ambient pressure, the saturation temperature T_s of steam, is 99·6°C, the specific volume v_v, is 1·694 m^3/kg. The values of specific internal energy and specific enthalpy for the liquid and vapour are also listed,

$$u_1 = 417 \text{ kJ/kg}, u_v = 2\,506 \text{ kJ/kg}; h_1 = 417 \text{ kJ/kg}, h_v = 2\,675 \text{ kJ/kg}.$$

The value of the latent specific internal energy and the latent specific enthalpy if not listed can easily be calculated from the data already obtained

$$u_{lv} = 2\,506 - 417 = 2\,089 \text{ kJ/kg}$$
$$h_{lv} = 2\,675 - 417 = 2\,258 \text{ kJ/kg}$$

The entropies s_1 and s_v are 1·303 kJ/kg K and 7·359 kJ/kg K, and s_{lv} if not listed can again be obtained by simple subtraction

$$s_{lv} = 7·359 - 1·303 = 6·056 \text{ kJ/kg K}$$

It will be observed that the difference between the values of u_v and of h_v is 169 kJ/kg. Equation (8.5) tells us that the small difference between them is equal to pv_v which from the tables is given by

$$pv_v = 1 \times 10^5 \times 1·694 \text{ (N/m}^2\text{) (m}^3\text{/kg)}$$
$$= 169·4 \text{ kJ/kg}$$

At this state point, defined by the pressure being 1×10^5 N/m^2, $T_s = 99·6$°C, when the fluid is dry saturated, we have,

$$u_v = 2\,506 \text{ kJ/kg (see Table 16.1)}$$
$$pv_v = 169·4 \text{ kJ/kg (from the above calculation)}$$
$$h_v = 2\,675 \text{ kJ/kg (see Table 16.1)}$$

The difference 169 given by the tables is only slightly different and well within the experimental error of the data from the value 169·4 given by equation (8. 5) which states, when written in the conventional symbols for energy per unit mass,

$$h = u + pv$$

(a) Saturated water and steam

p (N/m² × 10⁻⁵ or bars)	T_s (°C)	v_v (m³/kg)	u (kJ/kg) u_l	u_v	h (kJ/kg) h_l	hl_v	h_v	s (kJ/kg K) s_l	s_{lv}	s_v
0·10	45·8	14·67	192	2 437	192	2 392	2 584	0·649	7·500	8·149
1·00	99·6	1·694	417	2 506	417	2 258	2 675	1·303	6·056	7·359
200·00	365·7	0·00585	1 786	2 294	1 827	584	2 411	4·014	0·914	4·928

p	is pressure	u	is specific internal energy
T_s	is saturation temperature	h	is specific enthalpy
v	is specific volume	s	is specific entropy

The suffixes l and v signify saturated liquid and dry saturated vapour respectively.

(b) Superheated steam

p		Temperature (°C) 365·7	375	400	425	450	500
200	v × 10²	0·585	0·768	0·995	1·147	1·270	1·477
	h	2 411	2 605	2 819	2 955	3 062	3 239
	s	4·928	5·228	5·556	5·753	5·904	6·142

(c) Supercritical steam

p		Temperature (°C) 375	400	425	450	500
500	v × 10²	0·156	0·173	0·201	0·249	0·388
	h	1 717	1 879	2 064	2 288	2 722
	s	3·768	4·009	4·279	4·594	5·176

Quantities are taken from Thermodynamic and Transport Properties of Fluids by Y. R. Mayhew and G. F. C. Rogers

TABLE 16.1   Steam property tables

Use of property tables (Q and A)

Q. Find the saturation temperature of steam when the pressure is $0 \cdot 1 \times 10^5$ N/m
For saturated liquid at that pressure find the specific internal energy, enthalpy an
entropy. Find these also for the dry saturated state and also the specific volume.
For this pressure find the latent specific internal energy, enthalpy and entropy.

A. $T_S = 45 \cdot 8°C$

Liquid (sat.)	Vapour (dry sat.)	Differences
$u_l = 192$	$u_v = 2\,437$	$u_{lv} = 2\,245$ kJ/kg
$h_l = 192$	$h_v = 2\,584$	$h_{lv} = 2\,392$ kJ/kg
$s_l = 0 \cdot 649$	$s_v = 8 \cdot 149$	$s_{lv} = 7 \cdot 500$ kJ/kg K

16.6 Property tables—water, superheated and supercritical

When the temperature of steam is increased above the saturation ter
perature and is therefore represented by some point on the line EG of Figs. 16.1
and 16.2 the steam is said to be superheated. Properties of superheated vapour a
usually listed in a separate table from the properties of the saturated steam
although this depends on what property tables are used. However, from such a tab
the properties of superheated steam at, say, 200×10^5 N/m^2 can be obtained if the
temperature is also defined. Let us take for this example 400°C. Table 16.1 or
Ref. (3) show that the specific volume is $0 \cdot 00995$ m^3/kg, the specific enthalpy is
$2\,819$ kJ/kg and the specific entropy is $5 \cdot 556$ kJ/kg K. The saturation temperature
T_S of water at this pressure is 365·7°C, and from this one can determine that it ha
$(400-365 \cdot 7) = 34 \cdot 3$ degrees of superheat. The **Degrees of superheat** are the numbe
of degrees the temperature of the superheated steam is above the saturation temperature for that pressure.

Earlier in this chapter we mentioned the critical point and we can no
see how it can be recognised from the tables. In the column of the saturated table
(those discussed in section 16.6) giving the values of the saturation temperature t
highest value listed is 374·15°C. This is the **Critical temperature** and the pressur
corresponding to it is 221·2 bars and is called the **Critical pressure**. It can be see
from the properties listed that, at this temperature and pressure, $u_l = u_v$ and so u
is zero. In fact the two saturation lines in terms of internal energy have met.
Similarly in the same line $h_l = h_v$ and so h_{lv} is zero; the two saturation lines in
terms of h_l and h_v against T_S have met this point being known as the critical point
The critical point is the point where $u_{lv} = h_{lv} = s_{lv} \rightarrow 0$.

When steam is at a temperature higher than the critical temperature
it is said to be supercritical steam. Separate property tables are sometimes used
for supercritical steam. Consider an example; if the pressure is 550×10^5 N/m^2
and 450°C from the tables we can obtain values of the various properties—the
specific volume is $0 \cdot 00224$ m^3/kg, the specific enthalpy is $2\,227$ kJ/kg and the
specific entropy is $4 \cdot 494$ kJ/kg K.

Similar tables are available for ammonia, freon 12 and mercury.

Properties (Q and A)

Q. 1. What is the temperature of wet steam if its pressure is $1 \cdot 8 \times 10^5$ N/m²?
A. 1. The temperature of wet steam is the saturation temperature. In this case the tables give $T_s = 116 \cdot 9$°C.
Q. 2. What is the temperature of steam at $1 \cdot 8$ bars if its entropy is $1 \cdot 500$ kJ/kg K?
A. 2. The entropy shows that the steam is wet because $1 \cdot 500$ is greater than s_l ($1 \cdot 494$) and less than s_v ($7 \cdot 163$). Therefore the temperature is the saturation temperature $116 \cdot 9$°C.
Q. 3. Describe the state and evaluate the specific enthalpy of steam in the following conditions:

  (a) 130 bars, $330 \cdot 8$°C
  (b) 60 bars, 350°C
  (c) 400 bars, 600°C

A. 3.

  (a) Saturated—h is either 1 531 or 2 662 kJ/kg or somewhere between these
  (b) Superheated—h is 3 045 kJ/kg
  (c) Supercritical—h is 3 348 kJ/kg

Q. 4. 10 kg of water at 220 bars has an entropy of

  (a) $67 \cdot 42$ kJ/kg K
  (b) $45 \cdot 52$ kJ/kg K
  (c) $45 \cdot 00$ kJ/kg K
  (d) $42 \cdot 89$ kJ/kg K

What is its temperature? Describe its state in words.

A. 4.

  (a) 700°C—superheated steam
  (b) $373 \cdot 7$°C—dry saturated steam
  (c) $373 \cdot 7$°C—wet steam
  (d) $373 \cdot 7$°C—saturated water

Q. 5. Water at a pressure of 250×10^5 N/m² passes through a boiler. It is a liquid on entering and a vapour on leaving. What are the values of h_{lv} and s_{lv}?
A. 5. These are both zero as the pressure is above the critical pressure (follows a line similar to G_q in Fig. 16. 2).
Q. 6. What is the density of water in the dry saturated vapour and the saturated liquid states at 10×10^5 N/m²?
A. 6.

$$\rho_v = \frac{1}{v_v} = \frac{1}{0 \cdot 194} = 5 \cdot 16 \ \text{kg/m}^3$$

$$\rho_l = \frac{1}{v_l} = \frac{1}{0 \cdot 001128} = 887 \ \text{kg/m}^3$$

16.7 Summary

The state path of a system in terms of its temperature against time as energy is added to it has been followed from the solid to the gaseous state. The thermodynamic properties of saturated liquid, and of dry saturated, superheated, and supercritical steam and of water vapour have been examined with reference to the use of property tables. Other substances exhibit similar characteristics when their states are near the critical point or are those of wet vapour.

16.8 Questions for the reader

Q.1. With reference to the definition of solid and fluid given in this chapter classify the following substances when they are at normal temperatures and pressures:

(a)   Air
(b)   Water
(c)   Copper
(d)   Candle wax
(e)   Syrup
(f)   Mercury
(g)   Glass

(consider them under ambient conditions).

[f f s s f f s (sometimes considered a supercooled liquid)]

Q.2. What number of degrees of superheat has the steam under the following conditions:

(a)   At 1·0133 bar, 500°C
(b)   At 100 bar, 500°C ?

[400 degrees, 189 degrees]

Q.3. What saturation temperatures correspond to the following saturation pressures:

(a)   0·01 bar or 0·01 × 10^5 N/m^2
(b)   0·1 bar or 0·1 × 10^5 N/m^2
(c)   1·0 bar or 1·0 × 10^5 N/m^2
(d)   10·0 bar or 10 × 10^5 N/m^2
(e)   100 bar or 100 × 10^5 N/m^2 ?
[7, 45·8, 99·6, 179·9, 311·0°C]

Q. 4. What latent specific enthalpies and latent specific entropies correspond to the following saturation pressures:

(a) 2 bar
(b) 20 bar
(c) 200 bar
(d) 2000 bar?

$[h_{lv}$ (a) 2 202,   (b) 1 890,   (c) 584,   (d) —kJ/kg
s_{lv} (a) 5·597,   (b) 3·893,   (c) 0·914,   (d) —kJ/kg K]

Q. 5. Show that the relationship

$$h = u + pv$$

holds for dry saturated vapour at a pressure of 1×10^3 N/m^2.

[Yes, $h_v = 2\,514$ kJ/kg, $v_v = 129·2$ m^3/kg and $u_v = 2\,385$ kJ/kg]

Q. 6. For the ideal gas rule (equation (18. 1))

$$pV = RT$$

where R = 8·3143/18 kJ/kg K

Show whether this relationship holds for dry saturated vapour at 200 bar pressure.

[No, over 100 per cent out]

Q. 7. Calculate whether the equation holds for a pressure of 0·1 bar and a temperature of 500°C.

[Yes, within 2 per cent]

Q. 8. If you use a thermally insulated kettle for heating 0·1 kg of water, both kettle and contents being initially at 25°C and the atmospheric pressure being $1·0133 \times 10^5$ N/m^2, how much energy must you add to boil the water until the kettle is dry? While raising the temperature of kettle and contents to boiling point assume that $\frac{du}{dT}$ for the water is constant at 4 kJ/kg K and that the kettle requires 5 per cent of the energy the liquid requires during that period.

[269 kJ]

17

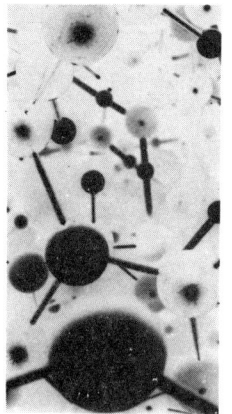

Properties of water

The properties of water, as a liquid, vapour and gas, are considered but what is said is not limited to water and is usually applicable to other fluids. Variations of temperature with pressure, pressure with volume, temperature with entropy, and enthalpy with entropy are used as practical illustrations of related properties.

17.1 Temperature-pressure diagram

The vertical axis of Fig. 17.1 represents absolute temperature in degrees K and the horizontal axis absolute pressure in $N/m^2 \times 10^{-5}$. We know from experiment that water is solid at a temperature of 273 K and below when its pressure is 1×10^5 N/m^2 (1 bar), and therefore we can draw a line of nearly constant temperature through 273 K and say that all points below the line represent ice, considering simply only one form of solid water. (In fact there are many forms of ice just as there are many forms of carbon—graphite, diamond, coal, etc.) Beginning at the point representing $0 \cdot 000611$ N/m^2 and 273 K we can, by taking values from the steam property tables, draw a curve of saturation temperature T_s against pressure. This is the saturation line shown in Fig. 17.1 to the right of which the fluid is water and to the left of which it is vapour. At the lowest point TP of the curve it will be seen that the solid, liquid and vapour states can all exist together.

At the point where the solid, liquid and vapour forms of water co-exist because we have the three components present, this is known as the **Triple point**, TP, and experimentally we know that at this point T = 273 K and p = $0 \cdot 00611 \times 10^5$ N/m^2. The water will be in a vapour state when represented by any state above the solid line, SS′, and to the left of the saturation line. All points representing states above the solid line, SS′, but to the right of the saturation line represent water in the liquid state. The saturation line ends at the critical point, CP, T = 647 K, p = 221 × 10^5 N/m^2 because above this there is no distinguishing between liquid and vapour. At supercritical conditions a system does not go through the wet vapour phase, such

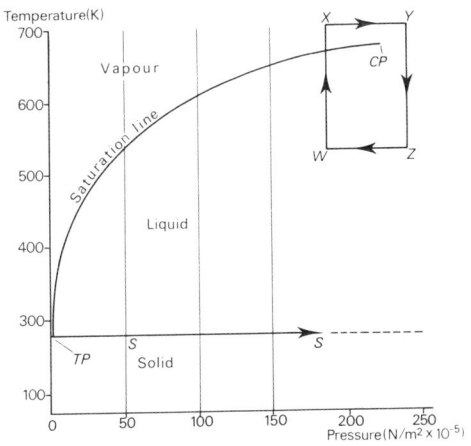

FIG 17.1   Temperature-pressure diagram for ice, water and steam

as that between D and E in Fig. 16. 2, as it changes directly from liquid to vapour. Although evaporation can be induced along the line WX in Fig. 17·1 or condensation along XW where WX or XW crosses the saturation line, a change between the same two states A and B can also occur along line WZYX or conversely along XYZW which does not cross the saturation line. In these latter cases neither goes through a phase of wet vapour. The changes from vapour to liquid along XYZW occurs simultaneously throughout the system, while the temperature and pressure drop from Y via Z to W. If the system is large it is practically difficult to maintain uniform pressure and temperature but this is a practical problem not a theoretical difficulty. Observation would show no moisture in the system at Y but, if conditions throughout the system are kept uniform, the bonds between molecules become stronger as the state changes from X to W via Y and Z, and the volume of the system becomes smaller during the transition from vapour at X to liquid at Z.

17.2 Pressure-volume diagram

In the property tables there is normally a column of figures representing pressure in $N/m^2 \times 10^{-5}$ (bars) and nearby a corresponding column for the specific volume of the dry saturated vapour in m^3/kg, from which the saturated vapour line of Fig. 17. 2 can be drawn. Values of the specific volume for saturated water in m^3/kg for various values of pressure will also be given, which can be used to plot a saturated water curve. Such curves are shown in Fig. 17. 2. Also shown in Fig. 17. 2 is a horizontal constant-pressure line C D E F G, which shows how specific volume changes during the heating process represented by C D E F G in Fig. 16. 1. As a substance changes from a solid to a gas when energy is added at constant pressure its density, $1/v$, decreases as is observed in Fig. 17. 2.

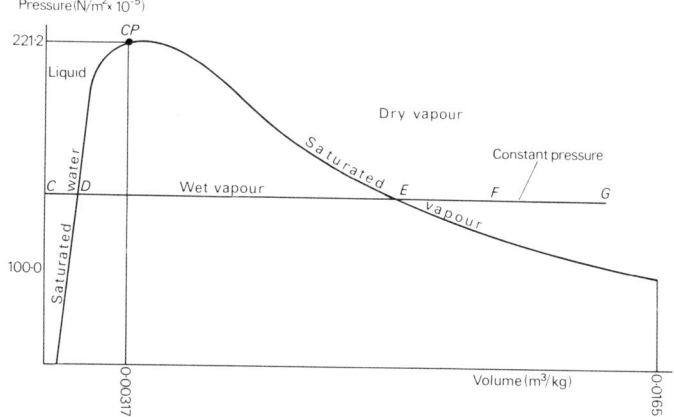

FIG 17.2   Pressure-volume diagram for water and steam

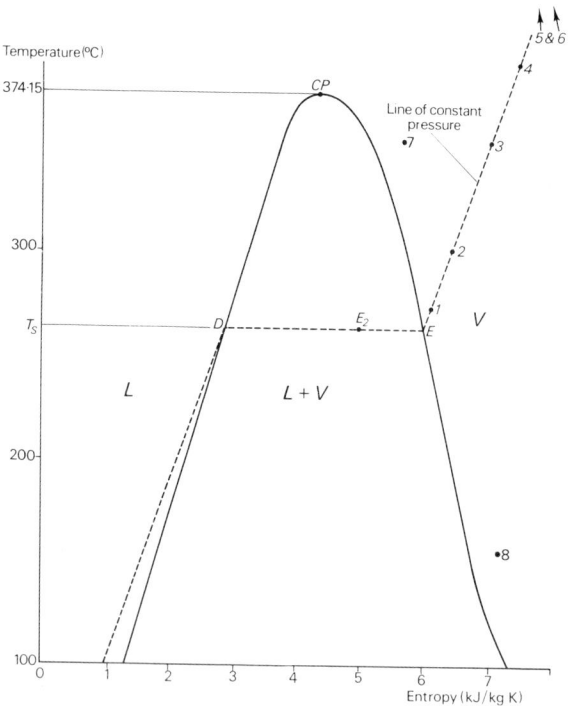

FIG 17.3   Temperature-entropy diagram for water and steam

17.3 Temperature–entropy diagram

The saturation temperature (see Chapter 16) and the specific entropy (defined in Chapter 14) of saturated liquid and of dry saturated vapour at that temperature are shown plotted on a temperature-entropy diagram in Fig. 17.3. The diagram can be used to show the values of entropy at various temperatures and pressures. One constant pressure line similar to that shown in Figs. 16.1 and 17.2 is shown in Fig. 17.3. It follows the constant pressure state path from a low temperature (too low to be shown on Fig. 17.3 where the solid has just melted) to respectively, saturated liquid, D; wet vapour, D-E, dry saturated vapour, E, and on finally to superheated vapour, F, eventually becoming an ideal gas, G, when sufficiently removed from the critical point to obey $pV = RT$.

The temperature-entropy diagram is frequently used to define complicated thermodynamic cycles in a way that will be described in Chapter 22.

17.4 Enthalphy–entropy diagrams

Entropy was defined in Chapter 14 particularly by equation (14.2) and enthalpy in Chapter 8 by equation (8.5). Values of these for liquid and vapour in the

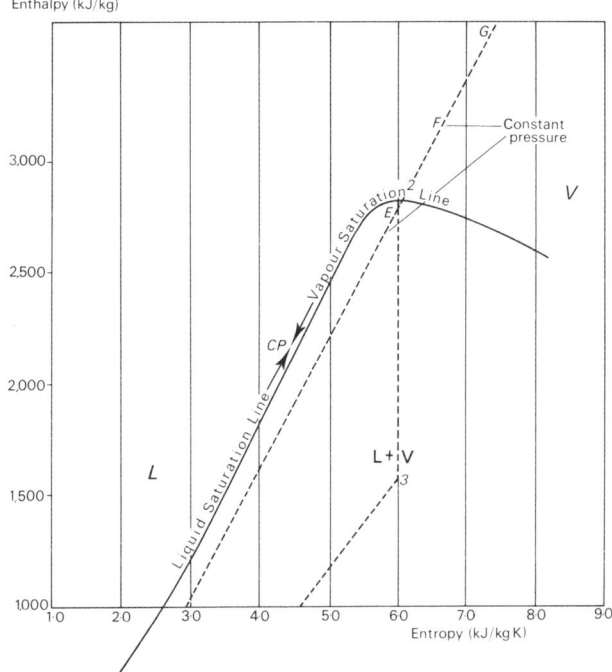

FIG 17.4   Enthalpy-entropy diagram for water and steam

saturated, superheated, and supercritical states are given in steam property tables. Taking the saturation values only, liquid and vapour saturation lines join at the critical point (see Chapter 16). This is plotted out in Fig. 17. 4. The constant pressure line originally shown in Fig. 16. 1 is also drawn but only the section above D leading to E F G is shown. This is because the scale is such that the solid and saturated liquid points C and D are far down below the axis of Fig. 17. 4—in fact off the page of the book! The dashed line from 2 to 3 will be discussed later.

To draw the whole of the constant pressure process A B C D E F G of Fig. 16. 1 on Fig. 17. 4 would be impractical because unless special scales were used either the figure would be very large or the scale would be too small to illustrate the wet vapour states.

Enthalpy-entropy diagrams for dryer vapours can be obtained and such a diagram, **Ref. (4)**, has been used freely in this book for the solution of problems. A part of Ref. (4) is shown in Fig. 25. 3.

T-s and h-s Diagrams (Q and A)

Q. 1.   Sketch a temperature-enthalpy diagram and show the state path of the working fluid of a steam power plant undergoing the following cycle of processes,

> 1-2   Heating at constant pressure in a boiler from saturated liquid to dry saturated vapour.
>
> 2-3   Isentropic expansion in a turbine, from the dry saturated state at boiler pressure, to the condenser pressure.
>
> 3-4   Cooling at constant pressure in a condenser.
>
> 4-1   Isentropic compression in a feed pump to a saturated liquid state in the boiler.

A. 1.   See Fig. 17. 5.

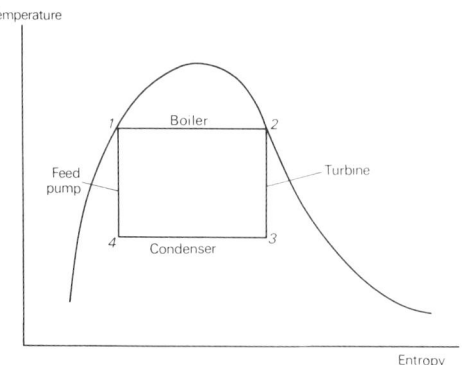

FIG 17. 5   Answer to question 1

Q. 2. Sketch an enthalpy-entropy diagram like that of fig. 17. 4 and, partially at least, sketch on it the cycle of processes referred to in Question 1.
A. 2. See Fig. 17. 6.

This part of the cycle has already been shown dashed in Fig. 17. 4.

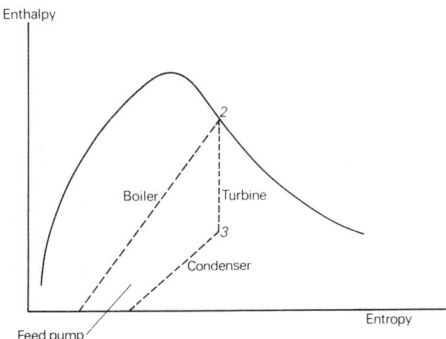

FIG 17. 6   Answer to question 2

Q. 3. Using values for enthalpy taken from an enthalpy-entropy diagram like that of Ref. (4), find the work done by the working fluid passing through a turbine as part of the cycle referred to in Question 1. Assume change of kinetic and gravitational energies are negligible. The boiler pressure is 20×10^5 N/m^2 and that of the condenser is 0.40×10^5 N/m^2. The turbine is isentropic.
A. 3. State point 2 is where the line of pressure 20×10^5 N/m^2 crosses the saturation line ($h_2 = 2\,800$ kJ/kg). State point 3 is on the constant entropy line through point 2 where that line crosses the line of pressure 0.4×10^5 N/m^2 ($h_3 = 2\,170$ kJ/kg).

From equation (9. 6)

$$
\begin{aligned}
-w_T &= \Delta h \\
&= -(h_3 - h_2) \\
&= -(2\,170 - 2\,800) \\
&= 630 \text{ kJ/kg}
\end{aligned}
$$

Q. 4. The boiler and condenser pressures of Question 1 are 10 and 0·1 bar. Find T for points 1, 2, 3 and 4, also find h for points 2 and 3.
A. 4. With regard to T as the conditions in the boiler are saturated $T_2 = T_1 = 180°C$ (found from tables). As regards the condenser both conditions are wet

vapour so,

$$T_3 = T_4 = 46°C \text{ (found from tables)}$$

vapour.   Values of h, h_2 can be found from the tables as it is dry saturated

$$h_2 = 2\,778 \text{ kJ/kg}$$

but h_3 must be found using the same technique as in Question 3:

$$h_3 = 2\,090 \text{ kJ/kg}$$

17.5 Dryness

At D in Fig. 17.3 the fluid was all liquid. That is to say its dryness was zero. At E it was all vapour and its dryness was 1. At some intermediate point E_2 on Fig. 17.3 the fluid would have a **Dryness** σ defined by

$$\sigma = \frac{\text{quantity of dry saturated vapour in the fluid at } E_2}{\text{total quantity of fluid}} \tag{17.1}$$

$$= \frac{m - m_1}{m} \tag{17.2}$$

where m and m_1 are the total mass and the mass of saturated liquid at E_2 respectively. While changing its state from D to E the fluid is taking energy at constant temperature T_s and because T_s is a saturation temperature, then the change from D to E is being carried out at constant pressure. So far as the fluid is concerned it is of no consequence whether the energy is transferred to it by heat or by work, reversibly or irreversibly. Suppose the energy were transferred reversibly by heat into the fluid then the quantity of energy taken in by the fluid as its state changes from D to E_2 is $(\Delta Q_r)_{D-E_2}$ or $T_s (\Delta S)_{D-E_2}$ (see equation (14.1))
One can rewrite equation (17.1) as follows

$$\sigma = \frac{\text{total quantity of energy transferred reversibly between D and } E_2}{\text{total quantity of energy transferred reversibly between D and E}}$$

The quantities in terms of mass have now been cancelled out as they are the same top and bottom of the equation, so

$$\sigma = \frac{(\Delta Q_r)_{D-E_2}}{(\Delta Q_r)_{D-E}}$$

$$= \frac{T_S \ (\Delta S)_{D-E2}}{T_S \ (\Delta S)_{D-E}}$$

$$= \frac{(\Delta S)_{D-E2}}{(\Delta S)_{D-E}}$$

$$= \frac{\text{length DE}_2}{\text{length DE}} \tag{17.3}$$

the lengths being measured on Fig. 17. 3. Equation (17. 3) gives, as required by the definition, a value of $\sigma = 1$ at point E and $\sigma = 0$ at point D.

The masses used in equation (17. 2) are shown in Fig. 17. 7 and, for any sample of wet vapour, are easily determined experimentally..

The dryness of wet steam is shown in Ref. (4). For example if the steam's specific enthalpy at 30×10^5 N/m^2 is 2 550 kJ/kg the dryness of the steam is 0·86.

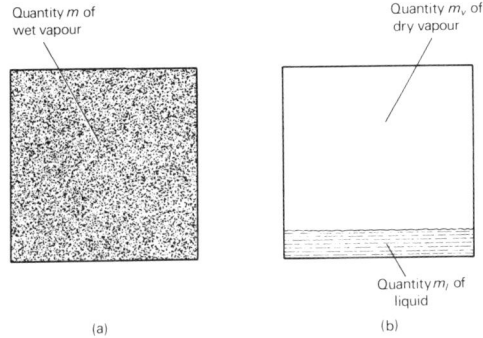

Quantity m of wet vapour

Quantity m_v of dry vapour

Quantity m_l of liquid

(a)

(b)

FIG 17. 7   Mixtures of liquid and dry vapours (a) wet vapour
(b) liquid and dry vapour

17.6 Summary

Many properties of water are given in property tables but they are often more usefully displayed in the form of charts. The most important of these is the enthalpy-entropy chart, sometimes called the Mollier chart after the man who invented it. Vertical isentropic lines are easily constructed and values of specific enthalpy can be read off and used in calculations. This chart can also be used for quickly determining the relationship between temperature, pressure, dryness, enthalpy and entropy of fluids near the critical point.

The whole thermodynamic cycle of water as a fluid and a vapour cannot be drawn on a Mollier chart unless the fluid remains as a wet vapour throughout the condensation or the diagram is on a very small scale. In order to show the whole cycle a temperature-entropy diagram is more useful.

Dryness has also been defined, as well as the triple point and the critical point.

17.7 Questions for the reader

Q. 1. Sketch the temperature-enthalpy diagram and mark approximately the places at which the state of each of the following is shown for a given constant pressure:

    (a) A gas
    (b) Dry saturated vapour
    (c) Superheated steam
    (d) Saturated liquid
    (e) Steam at 0·9 dryness

A. 1. See Fig. 17.8.

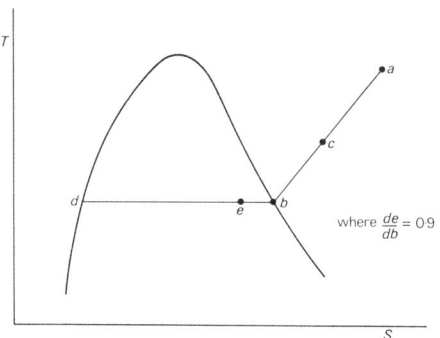

where $\frac{de}{db} = 0.9$

FIG 17.8

Q. 2. Repeat your answer to Question 1 on an enthalpy-entropy diagram. If part or all is impossible, state the reason for this.

A. 2. It is not practical to show point d because the chart would have to be very large. See Fig. 17.9

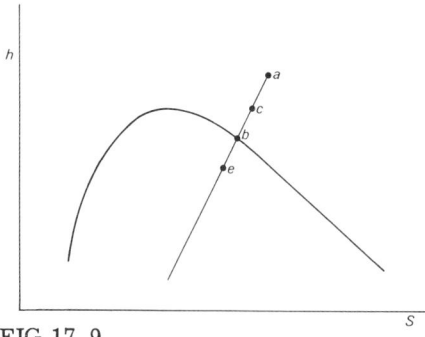

FIG 17.9

(In answering questions 3 to 8 use the Mollier diagram)

Q. 3. What is the specific enthalpy of steam at a pressure of 35×10^5 N/m^2 and at a temperature of 300°C? After an isentropic change of pressure to 0.8×10^5 N/m^2 what is its enthalpy? What are its entropies before and after the change?

$$[2\,980,\,2\,300\ \text{kJ/kg; both are } 6.45\ \text{kJ/kg K}]$$

Q. 4. If the specific enthalpy h of steam is $2\,500$ kJ/kg and its dryness σ is 0.90 what is its approximate temperature T and its pressure p?

$$[T = T_s = 128°C; p = 2.5 \times 10^5\ \text{N/m}^2]$$

Q. 5. If the specific entropy of steam is 7.2 kJ/kg K and its pressure 10×10^5 N/m^2 what is its state and its specific enthalpy h?

$$[\text{Superheated, h} = 3\,100\ \text{kJ/kg}]$$

Q. 6. If, for some steam with $\sigma = 0.80$, h = $2\,150$ kJ/kg what is p?

$$[p = 0.25 \times 10^5\ \text{N/m}^2]$$

Q. 7. If, for some steam, T = 400°C and h = $3\,250$ kJ/kg what are s and p?

$$[s = 7.18\ \text{kJ/kg K, p} = 18 \times 10^5\ \text{N/m}^2]$$

Q. 8. The specific entropy of some steam at 2×10^5 N/m^2 pressure, changes at constant pressure from s = 7.00 kJ/kg K by $\Delta s = -1.00$ kJ/kg K. What is its change of enthalpy?

$$[h_1 = 2\,660, h_2 = 2\,270; \Delta h = -390\ \text{kJ/kg}]$$

18 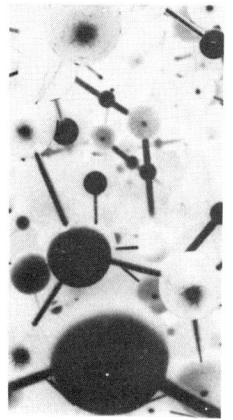 *Properties of a gas*

An ideal gas is defined in terms of the ideal gas rule, and the gas constant is expressed in terms of the specific enthalpy c_p and the specific internal energy c_v per degree. Two ideal gases, the perfect and the semi-perfect, are defined. The change of entropy of a perfect gas is evaluated for isentropic, polytropic and isothermal processes.

18.1 A gas and other fluids

It has already been shown in Fig. 16.1 that, if energy is transferred at constant pressure to a fluid in its liquid phase, its temperature rises from C until it reaches a point D at which the liquid is said to be in a saturated state. Point D is such that if the transfer to energy to the fluid continues at the same constant pressure the bonds between molecules are relaxed and the temperature will not rise until this relaxing of bonds has been completed and its state reaches point E on Fig. 16.1. Its state at point E is such that all the liquid has disappeared because the bonds between all the molecules have been sufficiently relaxed so that the fluid becomes a dry vapour. It is also called saturated because if the transfer of energy continues at the same constant pressure the vapour's temperature will rise to a value greater than that of dry saturated vapour and then the vapour is said to be superheated. While more energy is being transferred to the vapour its molecules move further apart and it becomes more like a gas if we define a gas as a fluid that obeys the ideal gas rule (see section 18.2). Eventually as its stored energy is increased the fluid's temperature rises to G in Fig. 16.1 where it is an ideal gas, obeying the ideal gas rule.

18.2 Definition of an ideal gas

Boyle's law states that at constant temperature the specific volume of an ideal gas varies inversely with the pressure or $(pv)_T = $ constant, in which p is the absolute pressure, v the specific volume and T the absolute temperature of the system. If that is so, the state path of an isothermal change shown on a pressure-volume diagram would be like that shown in Fig. 18. 1(a).

Charles' law states that at constant pressure the volume of an ideal gas varies linearly with temperature, or $(v/T)_p$ = constant, and so a constant pressure line on a temperature-volume diagram would be like that shown in Fig. 18. 1(b).

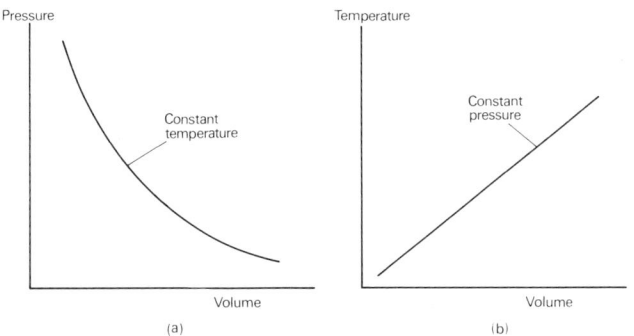

(a)

(b)

FIG 18. 1   Variation of volume with pressure and temperature

The two laws derived from experiment may be combined for, if we consider a system of unit mass, we know

from Boyle's law $(pv)_T$ = constant, and

from Charles' law $\left(\dfrac{v}{T}\right)_p$ = constant,

It follows that pv/T = constant

or $pv = RT$ (18. 1)

where R is a constant known as the **gas constant**. This is an equation that applies to a system of unit mass. When written for a system containing M units of mass occupying a volume V the equation becomes

$pV = MRT$ (18. 2)

where $V = Mv$.
Equation (18. 2) is known as the **Ideal gas rule** and any fluid that obeys this rule is defined as an ideal gas.

Equation (18. 2) may alternatively be derived from kinetic gas theory. This make two basic assumptions about the molecules,

(a)   The molecules occupy negligible volume
(b)   There are no forces between molecules.

It is assumption (b) in practice that fails first.

18.3 A gaseous system and its energy

In section 8.3 it was stated that at zero energy the volume of a system was taken to be effectively zero, but if energy were transferred into such a system the volume of the system would increase. In doing so the system would do work on its surroundings causing an energy transfer out of the system. If, when the heating stopped, the system did not collapse its volume would be maintained by virtue of the energy stored in the system. So there are two energy transfers and a change in stored energy to be considered,

(a)   The energy q transferred by heat, say, into a system of unit mass.
(b)   The energy w = pv transferred out by work from a system of unit mass against p, the pressure of the surroundings while the volume increases from 0 to v.
(c)   The energy u' stored in the system if q and w are not equal.

Assuming the system does not change its position, and therefore that Δk and Δz are both zero, equation (3.2) tells us that the relationship between q, w and u starting from absolute zero (in which case u would equal Δu) is

$$q - w = u$$

or $$q = u + pv \qquad (18.3)$$

18.4 Specific energies per degree

In this section the specific stored energies per degree rise in temperature, c_p and c_v, will be introduced. Sometimes they are called specific heat capacities at constant pressure and at constant volume respectively. Such names are misleading and should not be used. The change of enthalpy in the case of c_p and of internal energy in the case of c_v can be effected as well by work as by heat and therefore the words 'specific heat' could as well be 'specific work' although that would be equally wrong. In any case, as heat is an action that ceases to be when the action ceases, and as 'capacity' implies an ability to store, one is using incompatibles in using the words 'heat' and 'capacity' together.

The **Specific enthalpy per degree**, c_p, is defined as the rate of change of enthalpy, with respect to temperature, at constant pressure. This definition makes c_p the slope of the curve in Fig.18.2(a) where it is seen that

$$c_p = \frac{dh}{dT} \qquad (18.4)$$

The **Specific internal energy per degree**, c_v, is defined as the rate of change of internal energy, with respect to temperature, at constant volume. This definition makes c_v the slope of the curve in Fig. 18. 2(b) where it is seen that

$$c_v = \frac{du}{dT} \tag{18.5}$$

The use of the words 'at constant pressure' and 'at constant volume' is correct so far as the definitions based on Figs. 18. 2(a) and (b) are concerned but the words are confusing because c_p and c_v apply whenever there is a change of temperature whatever the conditions of pressure and temperature.

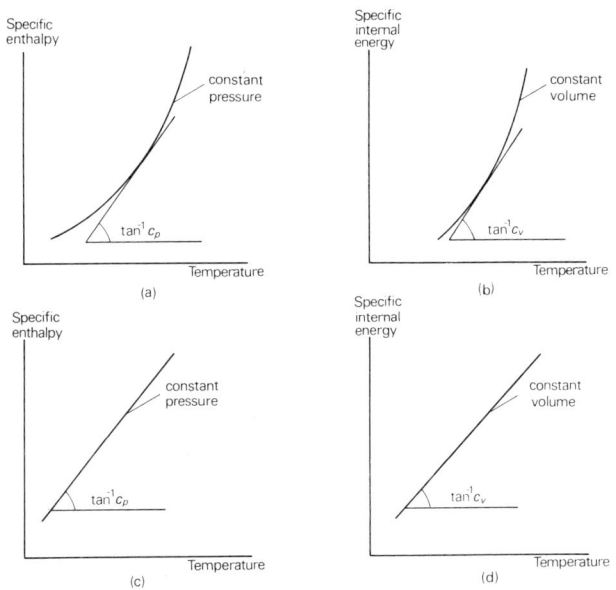

FIG 18. 2   Specific energies per degree

In section 8. 3 the enthalpy of a system was defined as stored energy equal in amount to the energy that would have to be transferred to the system, if the only means of inward transfer were by heat at constant pressure, to bring the system from a state of zero energy to its present state. This energy is the same as q in equation (18. 3) and therefore the enthalpy becomes

$$h = u + pv \tag{18.6}$$

If the enthalpy of the system changes with an associated rise in temperature, the changes of enthalpy and of temperature will be related, from equation (18.4), by

$$\Delta h = \frac{dh}{dT} \Delta T$$

$$= c_p \, \Delta T \tag{18.7}$$

where c_p is called the specific enthalpy per degree. In a **Perfect gas** c_p is by definition a constant in all conditions but for most real gases it changes with pressure and temperature. A perfect gas is therefore a gas that obeys the ideal gas rule and which in addition has a value of c_p that is constant for all temperatures and pressures. It will be seen from equation (18.9) that, if R and c_p are constant, c_v is also constant.

Equation (18.6) also had in it a term pv which was the energy transferred out of the system by work by reason of its change of shape—positive work is work done by a system on its surroundings. The stored internal energy u is not associated with change of shape. While the amount of u stored in a system is being increased the temperature rises and so the amount of energy stored in the system at constant volume can be written, in terms of the change of temperature, from equation (18.5),

$$\Delta u = \frac{du}{dT} \Delta T$$

$$= c_v \, \Delta T \tag{18.8}$$

where c_v is the specific internal energy per degree.

18.5 The gas constant

An ideal gas has been defined as a fluid that obeys the ideal gas rule, equation (18.2), or if applied to a system of unit mass, equation (18.1),

$$pv = RT \tag{18.1}$$

The definition of enthalpy as stated in equation (8.5) is, for a system of unit mass, equation (18.6)

$$h = u + pv \tag{18.6}$$

and from these equations we get

$$h = u + RT \tag{18.9}$$

which, if differentiated with respect to T gives

$$\frac{dh}{dT} = \frac{du}{dT} - R$$

$$\therefore R = \frac{dh}{dT} - \frac{du}{dT} \qquad\qquad (18.10)$$

$$= c_p - c_v \qquad\qquad (18.11)$$

from equations (18.4) and (18.5).

This constant is the gas constant for a particular gas and its value will vary from one substance to another. Values of c_p and c_v are listed in property tables such as those in Ref. (3).

At the beginning of this section we stated that the ideal gas rule for a system of unit mass could be written as,

$$pv = RT$$

Steam can be tested for its properties as a gas by taking several points 1, 2, 3, 4, 5 and 6 on a constant pressure line in Fig. 17.3 starting near the saturation line and getting successively nearer to the area in which we have said the fluid is a gas. Figure 17.3 is a temperature-entropy diagram and, if a line of constant pressure is considered, say 40×10^5 N/m^2, values of pressure, temperature and specific volume can be taken from the property tables for superheated steam. The values shown in Table 18.1 in this chapter for the six points 1-4 on the diagram of Fig. 17.3 and for 5 and 6 which are further along the constant pressure line outside the limits of this diagram have all been taken from property tables. It can be seen that the values of R approach more nearly a constant value the further one goes from the saturation line until the fluid obeys the ideal gas rule, confirming so far as steam is concerned, our earlier statement that the further one moves up the constant pressure line in the direction of increasing temperature the more nearly did steam move towards being an ideal gas.

All the points 1-6 in Table 18.1 are for a pressure of 40×10^5 N/m^2. If we keep near the saturation line but move nearer the critical point to point 7 on Fig. 17.3 and much further from it to point 8 we get values of R shown against points 7 and 8 in Table 18.1, the values of pressure and specific volume being taken

TABLE 18.1   Values of pv/T near and far from the critical point

Point	Pressure (N/m$^2 \times 10^{-5}$)	Temperature (C)	(K)	Volume (m^3/kg)	$R = \dfrac{pv}{T \times 10^3}$ (kJ/kg K)
1	40	260	533	0·0515	38×10^{-2}
2	40	300	573	0·0588	41×10^{-2}
3	40	350	623	0·0664	42×10^{-2}
4	40	400	673	0·0733	43×10^{-2}
5	40	500	773	0·0864	44×10^{-2}
6	40	600	873	0·0988	45×10^{-2}
7	160	350	623	0·00976	25×10^{-2}
8	10	200	473	0·2061	43×10^{-2}

from property tables. On the evidence of the three points 1, 7 and 8 it appears that the value of R becomes more constant as the state point is further from the critical point and in all cases this is found to be so.

18. 6 Two ideal gases

There are two ideal gases discussed in this book. The first one, a **Semi-perfect gas,** which we have already mentioned is a gas that obeys the ideal gas rule, equation (18. 2). This implies that R is a constant and, from equation (18. 11), this means that $(c_p - c_v)$ must be constant although neither c_p nor c_v individually need be constant. A **perfect gas,** however, must not only obey the ideal gas rule, R being a constant, but also c_p must be a constant. This of course means that c_v is constant too and for a perfect gas it follows that the constant pressure line in Fig. 18. 2(a) must be straight as it is in Fig. 18·2(c) and that the constant volume line in Fig. 18. 2(b) must also be straight as it is in Fig. 18. 2(d).

Summing up, for a semi-perfect gas, R in equation (18. 2) is a constant and for a perfect gas R and c_p are both constant where

$$R = c_p - c_v; c_p = \frac{dh}{dT} \text{ and } c_v = \frac{du}{dT}$$

18. 7 Entropy in a perfect gas

In equation (14. 2) an entropy change is defined in terms of reversible energy transfer by heat,

$$\Delta s = \int_1^2 \frac{dq_r}{T}$$

$$= \int_1^2 \frac{du}{T} + \int_1^2 \frac{dw}{T}$$

from equations (3. 2) and (8. 11) if the changes in kinetic and gravitational energies are negligible. Substituting for du and dw

$$\Delta s = \int_1^2 c_v \frac{dT}{T} + \int_1^2 \frac{p \, dv}{T}$$

from equations (18. 8) and (18. 1), and

$$\Delta s = \int_1^2 c_v \frac{dT}{T} + \int_1^2 R \frac{dv}{v}$$

from equation (8. 1). For a perfect gas c_v and R are constants

∴ $$\Delta s = c_v \log_e (T_2/T_1) + R \log_e (v_2/v_1)$$

Because $c_v = c_p - R$ from equation (18.11), it follows that

$$\Delta s = c_p \log_e (T_2/T_1) - R \log_e (T_2/T_1) + R \log_e (v_2/v_1)$$

$$= c_p \log_e (T_2/T_1) - R \log_e (P_2/P_1) \tag{18.12}$$

18.8 Reversible adiabatic and polytropic changes in a gas

In Chapter 13 we said that when considering thermodynamic cycles we are particularly interested in two processes. One was a simple work transfer process and the other a simple heat transfer process. The simple work transfer was adiabatic and the best simple work transfer was reversible and adiabatic, i.e. isentropic. In terms of equation (18.12) the best simple work transfer occurs when

$$\Delta s = 0$$

or

$$0 = c_p \log_e (T_2/T_1) - R \log_e (p_2/p_1)$$

$$0 = \gamma \log_e (p_2 v_2/p_1 v_1) - (\gamma - 1) \log_e (p_2/p_1)$$

where

$$\gamma = c_p/c_v$$

∴

$$\left(\frac{p_2 v_2}{p_1 v_1}\right)^\gamma = \left(\frac{p_2}{p_1}\right)^{\gamma - 1}$$

or

$$p_1 v_1^\gamma = p_2 v_2^\gamma$$

or

$$pv^\gamma = \text{constant} \tag{18.13}$$

Equation (18.13) gives a relationship between pressure p and volume v throughout an adiabatic and reversible—that is to say isentropic—change.

The change represented by equation (18.13) is the best adiabatic change for a perfect gas. As circumstances are such that real processes are not isentropic equation (18.13) then only becomes an approximation. It is usually more accurate to say that an actual change in a practical gas takes place in accordance with the rule

$$pv^n = \text{constant} \tag{18.14}$$

where n is a number lower than c_p/c_v and greater than one (see next section). A change of this kind is called a **Polytropic** change and each has its own polytropic index n depending on the real properties of the gas and the real circumstance in which the change occurs.

18.9 Isothermal change in a gas

Another special change in a perfect gas is a change that takes place at constant temperature. This, like the isentropic change, is really a special case of the polytropic change. In the isentropic case $n = \gamma$ and $pv^{\gamma} = \text{constant}$. The fact that $pv = \text{constant}$ in the isothermal case is consistent with the ideal gas rule, equation (18.1)

$$pv = RT$$

$$= \text{constant} \tag{18.15}$$

because R is a constant by definition and T is a constant in an isothermal change also by definition of the word isothermal.

18.10 Work done by gaseous systems

The steady-flow energy equation gives the useful work w_x done by a gas in steady flow. It is however often of interest to consider the work done by an enclosed mass of gas, for instance that enclosed in the cylinder of a reciprocating engine after the inlet valve has closed but before the exhaust valve opens. Such a system is shown inside a cylinder in Fig. 18.3. If the pressure exerted by the gas is p on a piston area A while it moves through a very small distance d, the work done is given by

$$W = \int pA\,dd \tag{18.16}$$

$$= \int p\,dv$$

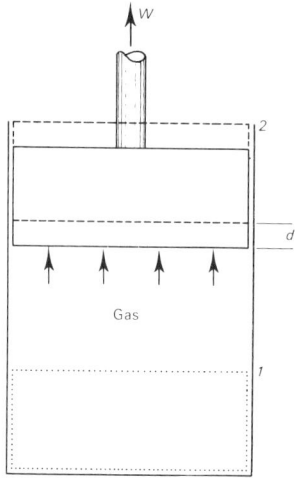

FIG 18.3   Gas doing work in a closed cylinder

If the expansion of the gas is isentropic

$$W = p_1 v_1{}^\gamma \int_1^2 \frac{dv}{v^\gamma} = \frac{p_1 v_1 - p_2 v_2}{\gamma - 1} \tag{18.17}$$

If the expansion is polytropic

$$W = p_1 v_1^n \int_1^2 \frac{dv}{v^n} = \frac{p_1 v_1 - p_2 v_2}{n - 1} \tag{18.18}$$

If the expansion is isothermal

$$W = p_1 v_1 \int_1^2 \frac{dv}{v} = p_1 v_1 \log_e (v_2/v_1) \tag{18.19}$$

All these expansions are shown on a p-v diagram in Fig. 18.4 where it will be seen that the work done, given by equation (18.17) is equal to the area under the curve.

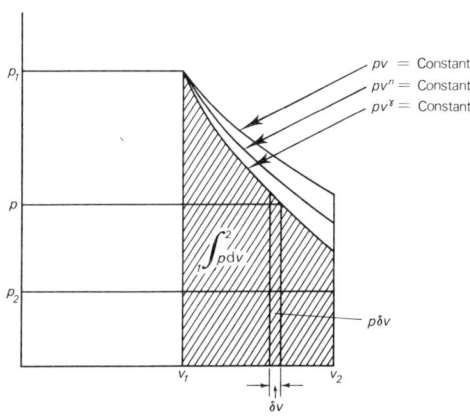

FIG 18.4   Isentropic, polytropic and isothermal changes of a gas on a p-v diagram

Perfect gases (Q and A)

Q.1.   A closed system of gas undergoes a change of volume from 17 mm³ to 39 mm³ starting from a pressure of 15×10^5 N/m². What is its final pressure if,

        (a)   The change is isentropic and $\gamma = 1 \cdot 4$
        (b)   The change is polytropic and $n = 1 \cdot 3$
        (c)   The change is isothermal?

A. 1.

$$p_1 = 15 \times 10^5 \text{ N/m}^2$$
$$v_1 = 1 \cdot 7 \times 10^{-8} \text{ m}^3$$
$$v_2 = 3 \cdot 9 \times 10^{-8} \text{ m}^3$$

(a) $p_2 = p_1 \left(\dfrac{v_1}{v_2}\right)^{\gamma} = 15 \times 10^5 \left(\dfrac{1 \cdot 7}{3 \cdot 9}\right)^{1 \cdot 4}$

$$= \frac{15 \times 10^5}{2 \cdot 3^{1 \cdot 4}}$$

$$= 4 \cdot 67 \times 10^5 \text{ N/m}^2$$

(b) $p_2 = p_1 \left(\dfrac{v_1}{v_2}\right)^n = \dfrac{15 \times 10^5}{2 \cdot 3^{1 \cdot 3}}$

$$= 5 \cdot 08 \times 10^5 \text{ N/m}^2$$

(c) $p_2 = p_1 \left(\dfrac{v_1}{v_2}\right) = \dfrac{15 \times 10^5}{2 \cdot 3}$

$$= 6 \cdot 52 \times 10^5 \text{ N/m}^2$$

Q. 2.   For each of the three changes in Question 1 find the work done by the gas.

A. 2.

(a) $W = \dfrac{p_1 v_1 - p_2 v_2}{\gamma - 1} = \dfrac{15 \times 17 - 4 \cdot 67 \times 39}{1 \cdot 4 - 1} \times 10^{-4}$

$$= 0 \cdot 0182 \text{ J}$$

(b) $W = \dfrac{p_1 v_1 - p_2 v_2}{n - 1} = \dfrac{15 \times 17 - 5 \cdot 08 \times 39}{1 \cdot 3 - 1} \times 10^{-4}$

$$= 0 \cdot 0190 \text{ J}$$

(c) $W = p_1 v_1 \log_e (v_2/v_1) = 15 \times 17 \times 10^{-4} \log 2 \cdot 3$

$$= 0 \cdot 0212 \text{ J}$$

18.11 Summary

The ideal gas rule states that $pV = MRT$ relating the pressure p and the volume V to the temperature T of a system of mass M, having a gas constant equal to R where $R = (c_p - c_v)$. The specific energies per degree, c_p and c_v, are defined in terms of rate of changes of enthalpy and internal energy per unit mass with respect to temperature. The change of entropy $\Delta s = s_2 - s_1$ between states 1 and 2 of a perfect gas is, in terms of temperature and pressure, given by the equation

$$\Delta s = c_p \log_e (T_2/T_1) - R \log_e (p_2/p_1)$$

The state paths and work done during an isentropic change were found to be

$$pv^\gamma = \text{constant and } W = \frac{p_1 v_1 - p_2 v_2}{\gamma - 1}$$

for an isentropic change where γ is the ratio c_p/c_v, and, for a polytropic change where n is the index,

$$pv^n = \text{constant and } W = \frac{p_1 v_1 - p_2 v_2}{n - 1}$$

(n depends on the gas and the conditions under which the change occurs). For an isothermal change

$$pv = \text{constant and } W = p_1 v_1 \log_e (v_2/v_1)$$

18.12 Questions for the reader

Q. 1. What is the temperature of a gas, for which the gas constant is 0·42 kJ/kg K, if the pressure is 10×10^5 N/m^2 and the specific volume is 0·020 m^3/kg ?

[47·6 K]

Q. 2. What is the volume of a quantity of gas, mass 40 kg, if its temperature is 400 K and its pressure is 2×10^5 N/m^2? The gas constant is 0·400 kJ/kg K.

[32 m^3]

Q. 3. What is the specific internal energy per degree of a gas for which R = 0·410 kJ/kg K and the specific enthalpy per degree is 0·700 kJ/kg K?

[0·290 kJ/kg K]

Q. 4. During a process the pressure and temperature of a perfect gas change from 1 000 K and 5×10^5 N/m^2 to 100 K and 1×10^5 N/m^2. By how much has the entropy changed? For specific energies per degree use the values from Question 3. Is the change of entropy affected by the process being carried out reversibly or not ?

[− 0·95 kJ/kg K; No, because entropy is a property.]

Q. 5. If a gas has the following properties at two state points is it in this range behaving as an ideal gas?

	Pressure (bar)	Volume (m³)	Temperature (°C)
A	15	0·132	200
B	30	0·099	400

⌊No⌋

Q. 6. If a gas has R = 0·189 kJ/kg K and c_V = 1.09 kJ/kg K what is the value of γ

[1·17]

Q. 7. In an isothermal process that occurs at 25°C the pressure of 10 kg of a perfect gas is increased from 10 bar to 100 bar. What is the change of entropy of the gas if c_p = 15 kJ/kg K and R = 4·16 kJ/kg K? If the process is reversible wha is the entropy change of the surrounding?

[− 95·8 kJ/K, + 95·8 kJ/K]

19

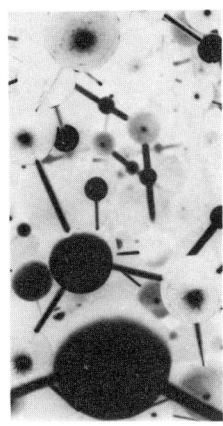

Mixture of perfect gases

For a pure substance in the gaseous phase there are two types of ideal gas, the perfect gas and the semi-perfect gas. In this chapter the behaviour of mixtures of ideal gases is considered and the gas constant, specific enthalpies and specific internal energies are derived from knowledge of the properties of the pure substances.

19.1 Avogadro's Law

Avogadro, an Italian scientist, stated in the early eighteenth century, but did not attempt to prove, that equal volumes of different gases at the same temperature and pressure contain the same number of molecules. This statement is called Avogadro's Law. We have accepted Avogadro's Law as true because it forms part of a consistent story that does not conflict with anything we observe.

Let us consider it with relation to the three systems (a), (b), and (c) of Fig. 19.1 in containers of equal volume V_0. Into each of the containers we have pumped sufficient of each of the three gases, all at the same temperature T_0, to bring the pressure in each of the three containers up to p_0. Because these three systems are each of the same volume, temperature and pressure, Avogadro's Law tells us that each contains the same number of molecules of its respective gas.

19.2 Atomic and molecular masses

There is a different gas in each of the three containers of Fig. 19.1. Because the containers have equal volumes and we pumped sufficient of each gas at the same temperature to the same final pressure in all three vessels then, assuming Avagadro's Law is true, we know that each vessel contains the same number of molecules.

Atoms of different gases are known to contain different numbers of protons, neutrons and electrons. Because of this the masses of different gases are

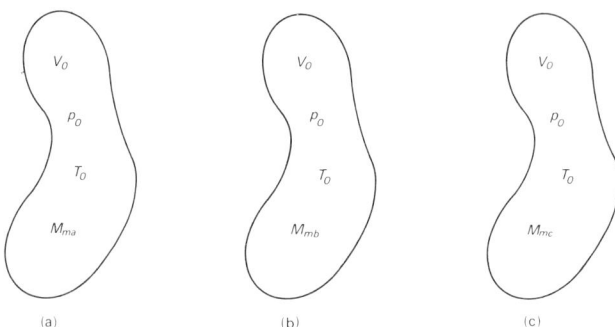

FIG 19.1   Three containers having same volume, each with a
different gas

different. The ratio of the mass of each gas atom to a hydrogen atom (the simplest
stable atom) is known as its **Atomic mass.** In fact we use an atom of hydrogen as a
unit of mass and we say hydrogen's atomic mass is 1·000. In more precise work
physicists and chemists have found it necessary to have more accurate values and
they say the atomic mass of hydrogen is 1·008. If instead of the atom we consider
the molecule, the molecular mass becomes important. The molecule is a number of
atoms forming a stable group larger than one atom—hydrogen in the gaseous state
on the whole forms molecules that comprise two atoms. So the molecular mass of
hydrogen H_2 is twice its atomic mass. The **Molecular mass** of a gas is the ratio
of the mass of each gas molecule to the mass of *an atom* of hydrogen.

In Fig.19.1 we have imagined things are so arranged that there is the
same number of molecules in each of the three containers, but because the gases of
(a), (b) and (c) are different gases they have different molecular masses M_{ma}, M_{mb}
and M_{mc}.

Let us suppose something about the mass of gas in container (a). Let
us suppose that its mass is in kg numerically equal to its molecular mass. That is
to say that the quantity of gas in container (a) is M_{ma} kg. There are the same
number of molecules of gas in containers (b) and (c) as there are in (a) and it
follows that the masses in kg in containers (b) and (c) must also be numerically
equal to their molecular masses M_{mb} and M_{mc} respectively.

A quantity of any gas equal numerically in kg to its own molecular
mass is called colloquially a **Mole**, or more correctly a kilogramme-mole. So there
is one mole of gas a in container (a), of gas b in (b) and of c in (c).

19.3 Universal gas constant

Each gas has its own gas constant R which is a constant for that gas.
There is a gas constant for the gas in container (a)—let us call this R_a—and for
those in containers (b) and (c); let us say R_b and R_c. We can write down the ideal

gas rule for each of the three gases in their containers

$$p_0 V_0 = M_{ma} R_a T_0$$
$$= M_{mb} R_b T_0$$
$$= M_{mc} R_c T_0$$

or, by rearranging the equations,

$$\frac{p_0 V_0}{T_0} = M_{ma} R_a$$
$$= M_{mb} R_b$$
$$= M_{mc} R_c$$
$$= R_0, \text{say}$$

R_0—a multiple of a gas's molecular mass and its gas constant—is therefore the same for all gases, there being no restriction on the number of gases to which the above argument can be applied. Therefore R_0 is called the **Universal gas constant**. In S.I. units it is expressed in kJ/kg-mol K. It is the same for every gas and has been found experimentally to be 8·3143 kJ/kg-mol J. Therefore, if one knows the molecular mass M_m of a gas, one can find the gas constant R in kJ/kg K because

$$R = \frac{R_0}{M_m} \tag{19.1}$$

19.4 Mixtures

Gases when mixed can become chemically stable or unstable. For instance oxygen and hydrogen form a stable mixture and, much to mankind's relief, do not react unless ignited. However gaseous fuel and oxygen under pressure in the cylinder of a reciprocating engine could well form a chemically unstable system as in a diesel engine. Mixtures in conditions of instability are discussed in Chapter 20 In this chapter we are concerned with mixtures in conditions of chemical stability.

19.5 Partial pressures, volumes and masses

A statement made by Dalton (1766-1844) led to a general acceptance of what is now known as the **Law of partial pressures**. This law states that the total pressure of a mixture of chemically stable gases is equal to the sum of all the individual pressures of the constituent gases if each occupied the space alone at the same temperature. This law can be written in the form

$$p_T = p_1 + p_2 + p_3 \tag{19.2}$$

where p_1, p_2 and p_3 are what the pressures of the three individual gases would be at, say, temperature T_T, if each alone occupied the total volume V_T; p_T is the total pressure of the mixture occupying volume V_T at T_T.

Suppose we consider the three gases to occupy separate containers so that gas 1 is in container 1, 2 in 2, and 3 in 3 as shown in Fig. 19. 2(a). The pres sures of the gases in their separate containers are p_1', p_2' and p_3', their temperatur are all T_T and this temperature is kept constant throughout the whole of the follow ing procedure. Now let the volume of containers 2 and 3 change from V_2' to V_2'' and from V_3' to V_3'' respectively the change of state being shown in Fig. 19. 2(b).

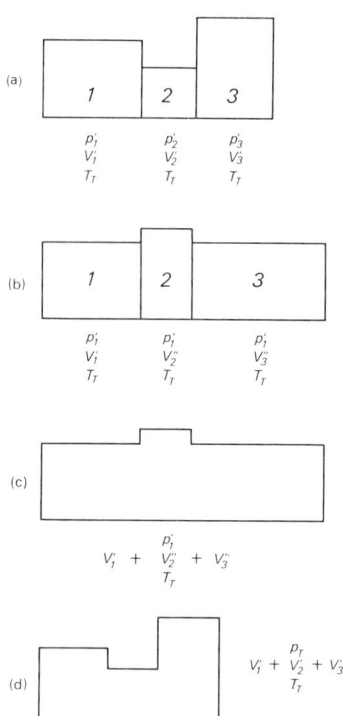

FIG 19. 2   Mixing gases

In this change we have assumed that the temperature has remained constant and the contents of containers 2 and 3 have not changed and so the pres sures of gases 2 and 3 will have changed. It has been arranged that the volume changes occur so that in the final state all three gases have the pressure p_1'.

As this is an isothermal change we can state, from equation (18.15) that,

$$V_2'' = \frac{p_2' V_2'}{p_1'} \qquad (19.3a)$$

and

$$V_3'' = \frac{p_3' V_3'}{p_1'} \qquad (19.3b)$$

Now let large holes be made in the partitions between containers 1 and 2, and 2 and 3, as shown in Figs 19.2(c) and (d). As all the gases are stable and at the same pressure in Fig. 19.2(c) they will diffuse into one another. The mixed gases will be at a pressure p_1' and will occupy the volume $(V_1' + V_2'' + V_3'')$. Now consider Fig. 19.2(d) where the volumes of containers 2 and 3 are returned to their original volumes V_2' and V_3'. They will then have a new pressure p_T and from equation (18.15)

$$p_1 (V_1' + V_2'' + V_3'') = p_T (V_1' + V_2' + V_3')$$

From this and equations 19.3(a) and (b).

$$p_T = \frac{p_1 (V_1' + V_2'' + V_3'')}{V_1' + V_2' + V_3'}$$

$$= \frac{p_1' V_1' + p_2' V_2' + p_3' V_3'}{V_1' + V_2' + V_3'}$$

$$= \frac{p_1' V_1' + p_2' V_2' + p_3' V_3'}{V_T} \qquad (19.4)$$

If the gas, originally in V_1, alone filled the total volume V_T (where $V_T = V_1' + V_2' + V_2' + V_3'$) the pressure would have been, let us say, p_1, and from equation (18.15)

$$p_1 V_T = p_1' V_1'$$

Similarly $p_2 V_T = p_2' V_2'$ and $p_3 V_T = p_3' V_3'$.

Hence from these and equation (19.4)

$$p_T = p_1 + p_2 + p_3 \qquad (19.5)$$

This was first demonstrated by Regnaut. Remember that the argument that led to equation (19.5) was based on the assumption that each gas in the mixture behaves as if it occupies the whole volume alone at its partial pressure and the common mixture temperature. So that if the volume of the constituent gases of a mixture occupying volume V_T are V_1, V_2 and V_3, then

$$V_T = V_1 = V_2 = V_3 \tag{19.6}$$

From an elementary consideration of the conservation of mass, the total mass of the mixture, M_T, is equal to the sum of the masses of the constituents, or

$$M_T = M_1 + M_2 + M_3 \tag{19.7}$$

19.6 Specific energies per degree

The total enthalpy in a stable mixture of perfect gases is conserved, s

$$M_T h_T = M_1 h_1 + M_2 h_2 + M_3 h_3$$

where h_T, h_1, h_2 and h_3 are the enthalpies of the mixture and of the constituent parts. Differentiating this equation with respect to temperature and dividing through by M_T we get

$$\frac{dh_T}{dT} = \frac{M_1 dh_1}{M_T dT} + \frac{M_2 dh_2}{M_T dT} + \frac{M_3 dh_3}{M_T dT}$$

or

$$c_{pT} = \frac{M_1}{M_T} c_{p1} + \frac{M_2}{M_T} c_{p2} + \frac{M_3}{M_T} c_{p3} \tag{19.8}$$

where c_{pT}, c_{p1}, etc., are the specific enthalpies per degree of the mixture and of the constituent parts.

Similarly it can be shown that

$$c_{vT} = \frac{M_1}{M_T} c_{v1} + \frac{M_2}{M_T} c_{v2} + \frac{M_3}{M_T} c_{v3} \tag{19.9}$$

From equations (18.11), (19.8) and (19.9)

$$R_T = c_{pT} - c_{vT}$$

$$= \frac{M_1}{M_T} R_1 + \frac{M_2}{M_T} R_2 + \frac{M_3}{M_T} R_3 \tag{19.10}$$

and for the mixture the ideal gas rule becomes

$$p_T V_T = M_T R_T T_T$$

for which the relationship given by equation (18.1) still holds for the mixture as it does for any of the components.

$$R_0 = R_T M_{MT} = R_1 M_{M1} = R_2 M_{M2}, \text{etc.}$$

It follows from equations (19.8) and (19.9) that a similar equation can be derived for the isentropic index, $\gamma = c_p \text{-} c_v$, relating the index γ_T for the mixture to the indices, γ_1, γ_2, etc., for the constituents.

Mixtures (Q and A)

Q. 1.   The atomic composition of a molecule of water is H_2O. The atomic mass of hydrogen is 1 and of oxygen is 16. If water were in a gaseous state what would be its gas constant?

A. 1.   This is not a mixture of gases but merely one gas, so we need only to apply equation (19.1),

$$R = \frac{R_0}{M_M} = \frac{8 \cdot 3143}{18} = 0 \cdot 462 \text{ kJ/kg K}$$

Q. 2.   A stable mixture is composed of oxygen, hydrogen and carbon dioxide in proportions by mass $3:2:1$. What is the gas constant and the isentropic index of the mixture at 300°K?

A. 2.

Constituent	M_M	R (kJ/kg K)	c_p* (kJ/kg K)	**c_v† (kJ/kg K)	γ‡
O_2	$16 \times 2 = 32$	$\frac{8 \cdot 3143}{32} = 0 \cdot 260$	0·918	0·658	1·395
H_2	$1 \times 2 = 2$	$= 4 \cdot 157$	14·310	10·153	1·409
CO_2	$12 + 32 = 44$	$= 0 \cdot 189$	0·846	0·657	1·288

* From tables, Ref. (3)
† $c_v = c_p - R$
‡ $\gamma = c_p / c_v$

Total mass $M_T = $ say, $3 + 2 + 1 = 6$ kg

Then $R_T = \frac{3}{6} (0 \cdot 260) + \frac{2}{6} (4 \cdot 157) + \frac{1}{6} (0 \cdot 189) = 1 \cdot 547 \text{ kJ/kg K}$

And $\gamma_T = \frac{3}{6} (1 \cdot 395) + \frac{2}{6} (1 \cdot 409) + \frac{1}{6} (1 \cdot 288) = 1 \cdot 382$

Q. 3.   A mixture of 6 kg of carbon dioxide gas, CO_2, 1 kg of hydrogen gas, H_2, and 2 kg of oxygen gas, O_2, are mixed at 300 K. What are the partial pressures of these gases if the mixture occupies 4 m³? Add these pressures together and compare the result with the result obtained by finding a gas constant for the mixture.

A. 3.

Gas	M_M	R (kJ/kg K)	p (N/m² × 10⁻⁵)
CO_2	$12 + 16 \times 2 = 44$	$\dfrac{8 \cdot 314*}{44} = 0 \cdot 189$	$\dfrac{0 \cdot 189 \times 300}{4} \times 6 \times 10^{-2} = 0 \cdot 851$
H_2	$1 \times 2 = 2$	$4 \cdot 157$	$3 \cdot 118$
O_2	$16 \times 2 = 32$	$0 \cdot 260$	$0 \cdot 390$

* from equation (19. 1)

$p_T = 0 \cdot 851 + 3 \cdot 118 + 0 \cdot 390 = 4 \cdot 359$ bars
R for mixture (from equation 19. 10) $= 0 \cdot 646$
p_T for mixture (from equation 18. 2) $= 4 \cdot 360$—cf. p_T above.

19. 7 Summary

The universal gas constant R_0 which is the same for all gases is the product of the gas constant for that gas and its molecular mass. The specific enthalpy per degree, c_{pT}, for a chemically stable mixture of gases has been shown to be given by,

$$c_{pT} = \frac{M_1}{M_T} c_{p1} + \frac{M_2}{M_T} c_{p2} + \frac{M_3}{M_T} c_{p3}$$

where c_{pT}, c_{p1}, etc., and M_T, M_1, etc., are the specific enthalpies per degree and the masses of the mixture and of the constituents respectively. Relationships between $p, v, c_p, \ldots$, etc., can be shown to apply to stable mixtures of gases as to individual gases by using Avogadro's law. Similar expressions can be written for the gas constant, the specific internal energy per degree and the isentropic index. An expression for the molecular mass of the mixture is shown.

19. 8 Questions for the reader

Q. 1. Given the atomic masses H = 1,   C = 12,   N = 14,   O = 16,   Cl = 35

what are the molecular masses of the following:

(a) Methane, CH_4
(b) Ethyl alcohol, C_2H_5OH
(c) Benzene, C_6H_6
(d) Chloroform, $CHCl_3$

[16, 46, 78, 119·5]

Q. 2. Find the volume of 1 kg of methane gas CH_4 at a temperature of 500 K and at a pressure of $2 \cdot 5 \times 10^5$ N/m^2.

[$1 \cdot 039$ m^3]

Q. 3. Find the gas constant for ethylene C_2H_4.

[$0 \cdot 297$ kJ/kg K]

Q. 4. Find the temperature of 5 kg of hydrogen H_2 in a gas bottle of volume $0 \cdot 5$ m^3 at a pressure of 15×10^5 N/m^2.

[$36 \cdot 1$ K]

Q. 5. For air with a mass analysis of $N_2 : O_2 : : 0 \cdot 767 : 0 \cdot 233$, calculate γ at 277°C and the mean molecular mass of the mixture.

[$1 \cdot 38, 29 \cdot 0$]

Q. 6. In a 1 m^3 container there is 1 kg of argon (Ar) and 1 kg of carbon dioxide (CO_2) at 27°C. What are the partial pressures of the components and the total pressure? (the molecular mass of Ar is 40 and of CO_2, 44)

[$0 \cdot 624, 0 \cdot 567; 1 \cdot 191$ bar]

Q. 7. If 20 kg of nitrogen and 1 kg of hydrogen are in a container at 127°C such that the total pressure is 10 bar, what is the volume of the vessel?

[$4 \cdot 04$ m^3]

Q. 8. A 50/50 mixture by mass of methane CH_4 and carbon monoxide CO is in a container at 400 K. Calculate c_{VT} and M_T for the mixture

[$1 \cdot 787$ kJ/kg K, $20 \cdot 4$]

20 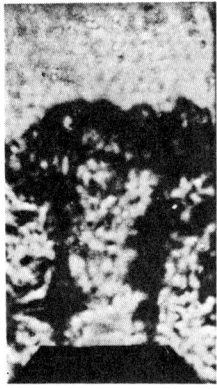 *Combustion*

In many practical heat engines part of the cycle includes combustion and this chapter deals with the elementary details of the combustion process. This process is considered for both a closed and flowing system and the energy release during the process is examined.

20.1 Reaction

To appreciate the role of combustion in engineering it is necessary to have some knowledge of elementary chemistry. This chapter cannot and does not attempt to cover rigorously even a small part of the science of chemistry. As was suggested in the opening sentences above it gives no more than an outline of some of the problems involved.

During the course of some reactions in order to maintain a constant temperature a continuous supply of energy to the reactants is required. Such reactions are called **ENDOTHERMIC** and do not concern us here because they do not provide a usable energy source. Other reactions release energy and are known as **EXOTHERMIC**. These are the reactions with which we are mostly concerned because energy release can be used usefully. The rate at which reactions can occur varies from extremely slow to very fast. They can occur between many pair of substances, and indeed more complex reactions can involve more than a pair of substances, but, because air is abundant and because one of its major constituents is oxygen, engineers are mostly concerned with rapid exothermic reactions between one substance, which he calls a **FUEL** and oxygen or substance containing oxygen. This reaction he calls **COMBUSTION** and when it occurs sufficiently rapidly he can use it as a source of energy. Fuels can be either fluids—gaseous or liquid—or solids and the same elementary principles can be applied to all these types of fuels. Typical compounds that can be used as fuels are propane, butane and octane, all of which contain carbon C and hydrogen H which combine with oxygen to form carbon monoxide CO or carbon dioxide CO_2 and water H_2O.

A source of energy that is now a major alternative to exothermic reactions is nuclear reaction which is briefly described in the next chapter.

20.2 The processes of combustion

An engineer must be able to answer the following questions about a fuel:

(a) Will it react with the oxidant (a compound that will supply oxygen)?
(b) If it reacts, under what circumstances does the reaction occur?
(c) What energy will be released during the reaction?

The simple answer, so far as fuel F_1 is concerned, is the following directional equation,

$$F_1 + O \longrightarrow C + A + energy \qquad (20.1)$$

which states that 'F_1 reacts with O and together they form products C and A (usually carbon dioxide and water)'.

It is useful to consider separately what happens if the mixture of reactants prior to, during, or after combustion,

(a) are in a state of steady flow,
(b) are contained in a closed vessel.

The nature of combustion is common to both situations.

20.3 Combustion in steady flow

Combustion in steady flow takes place in the furnace of a boiler, or in the combustion chamber of a gas turbine and in many other circumstances. There is a flow of fuel and oxidant into the combustion chamber. The flame, once ignited, continues to burn if a steady flow of a controlled mixture of fuel and oxidant enters the combustion chamber, replacing the products of combustion that are streaming away from the flame to the flue. Using the gas burner as an example of combustion in steady flow the processes of burning can be observed to take place in the following stages:

(a) The molecules of oxygen in the air are brought into close proximity with the molecules of fuel so that there is physical inter-action. If the apparatus has been well-designed this mixing will be very thorough.
(b) The temperature of the mixture rises due to increases in the mixture's pressure and also because of energy it receives by radiation from the flame.

So while the mixture approaches the flame front its temperature is increasing until it reaches the temperature at which the reaction occurs—called the **Ignition temperature.**

(c) During the reaction the oxygen and hydrocarbon (fuel) molecules can be thought of as dividing into atoms of oxygen, hydrogen and carbon which recombine to form water and carbon dioxide molecules. This breaking down of old bonds and the forming of new bonds for exothermic reactions releases energy that causes a further rise in temperature of the new mixture. This is often apparent by the mixture's becoming incandescent and taking on the appearance of what we call a flame.

(d) The new mixture is called the products of combustion and consists of some residual oxygen and all the nitrogen from the air with the newly formed water and carbon dioxide. There are also other gases in smaller quantities that come from the original air and fuel.

Because the changes are taking place in steady flow the steady flow energy equation (8.6) is applicable,

$$Q - W_X = \Delta H + \Delta K + \Delta Z \qquad (8.6)$$

or, if changes of kinetic and gravitational energy are negligible

$$Q - W_X = \Delta H \qquad (20.2)$$

20.4 Combustion in a closed system

The principles of combustion of a closed system are the same principles as those of steady flow with one important exception. This is that, whereas combustion in steady flow is usually taking place while the oxygen and fuel mixture is coming to and the products of combustion are going from the flame front, combustion of a closed system is carried out by a discrete quantity of the mixture in a vessel usually constant in size and shape. It takes place, say, in the cylinder of a reciprocating internal combustion engine as described in section 23.1. In such a cylinder, if the reaction occurs within a short period of time, the situation is similar to that of a system of the mixture within a vessel of constant, or nearly constant, size and shape. Within the cylinder a device, such as a sparking plug in the case of a petrol engine, raises the temperature of a small part of the mixture to the ignition temperature. The energy released by this part of the mixture on reaction raises the temperature of the rest of the mixture so that all the contents of the cylinder combust and consequently the temperature of the whole system is greatly increased above the ignition temperature. The beginning of a reaction is called **Ignition.** The combustion of the whole system mentioned above happens so quickly that, during its occurrence, the movement of the piston and the consequent change in volume of the cylinder are, to a first approximation, negligible and the combustion can be thought of as having happened at constant volume. (With increased engine speeds for high performance engines this statement may be challenged with increasing justification.) Because in this example one system only is involved, not a steady flow of systems, equation (3.2) applies,

$$Q - W = \Delta U + \Delta K + \Delta Z \qquad (3.2)$$

or if changes of kinetic and gravitational energy are negligible,

$$Q - W = \Delta U \qquad\qquad (20.3)$$

Knowledge that the combustion occurs in a closed system is part of the answer to the second question of section 20.2. Another part of the answer lies in the knowledge of the flash points of the mixture. The **Flash point** is the lowest temperature at which a reactant gives off sufficient vapour to form an inflammable mixture with air (under conditions of a standard test which conditions are the subject test specifications). The energy which must be added to a mixture of fuel and oxidant to cause combustion varies with the air/fuel ratio—this is shown in Fig.20.1.

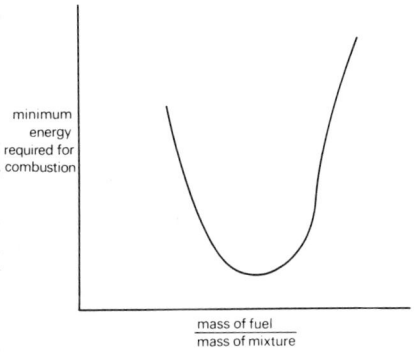

FIG 20.1   Changes of minimum energy for ignition with changes of mixture strength

20.5 The energy of combustion

Now we must answer the third of the three questions asked in section 20.2. If the fuel reacts 'What energy will be released during the reaction?'

The energy released has come from the internal energy and is part of the energy U of the fuel and oxidant. If the release of this energy takes place adiabatically (Q = 0) in a closed system at constant volume the energy is not released at all. It remains part of the internal energy U but in a different form. Whereas before combustion this energy was stored in the form of bonding between the atoms, now it is stored in the form of new bonding and of increased movement, apparent as a rise of temperature. It is all still called internal energy. Therefore, for a system undergoing an adiabatic combustion process in a constant volume, making $W = 0$, equation (20.3) becomes

$$\Delta U = 0$$

but we must remember that one should not infer from this that there has been no change. There has been a change from one form of internal energy to another which has resulted in a temperature rise. Such a change could not of course take place in a perfect gas because we have already defined a perfect gas as being one in which not only does $pV = MRT$ but also c_p and therefore c_v are constant. If c_v is constant, then,

$$c_v = \frac{dU}{dT}$$

from equation (18.5), or

$$\Delta U = c_v \Delta T$$

If U remains constant, therefore, in a perfect gas so also must T. T cannot increase in a perfect gas without an increase in U. The above equation is apparently inconsistent with the statement that some part of U can change from being in the form of bonding to being in the form of increased movement of the molecules, made apparent by a rise in temperature. The reactants may be perfect gases, as may the products but a change of chemical composition occurs when reaction occurs which means that $(c_v)_{mean}$ for the reactants is not necessarily equal to $(c_v)_{mean}$ for the products. An example of a perfect gas mixture as reactant and a perfect gas as product is

$$2H_2 + O_2 \longrightarrow 2H_2O + \text{energy}$$

if this is carried out under conditions where the water behaves as a perfect gas.

Let us now consider a mixture of fuel and oxygen inside a rigid vessel of constant volume—Fig. 20.2(a). The system's initial temperature T_0 being the same as that of its surroundings. When combustion takes place the temperature rises to T_1—Fig. 20.2(b)—which is higher than the ignition temperature but because the vessel is rigid and of constant volume no work is done. The vessel is not, however, thermally insulated and the heat energy Q is transferred out of the vessel. As a result of the heat transfer the temperature T_1 of the products falls to T and finally to the same temperature T_0 as that of the reactants before combustion. This final situation is the one shown in Fig. 20.2(c). Therefore from the first law, equation 8.3,

$$Q = \Delta U_0$$

$$= (U_{PO} - U_{RO}) \tag{20.4}$$

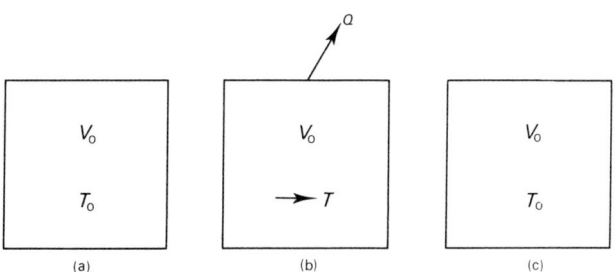

FIG 20. 2   Internal energy of combustion at reference temperature

where U_{P0} and U_{R0} are the internal energies of the products and the reactants respectively after and before the combustion and heat transfer.

The overall change of internal energy from the condition of Fig. 20. 2(a) to that of 20. 2(c) is $(U_{P0}-U_{R0})$ as stated in equation (20. 4). In conditions of steady flow or if the process occurs at constant pressure equation (20. 4) would become,

$$Q = \Delta H_0$$

$$= (H_{P0} - H_{R0}) \qquad (20. 5)$$

where H_{P0} and H_{R0} are the enthalpies of the products and of the reactants respectively after and before the combustion and the heat transfer. The energy ΔU_0 of equation (20. 4) and ΔH_0 of equation (20. 5) are called respectively the internal energy and the enthalpy of combustion at temperature T_0. These values are related by the equation

$$\Delta H_0 = \Delta U_0 + (p_{P0}V_{P0} - p_{R0}V_{R0})$$

which is an adaption of equation (8. 5). p_{P0}, V_{P0}, p_{R0} and V_{R0} are the pressure and volume of the products, and the pressure and volume of the reactants respectively, both the products and the reactants being at a temperature T_0. This equation is true for any state of the different products and reactants.

If, however, the products and reactants are composed of perfect gases we can write

$$P_{P0}V_{P0} = n_P R_0 T_{P0}; \qquad P_{R0}V_{R0} = n_R R_0 T_{P0} \qquad (20. 6)$$

where R_0 is the universal gas constant, and n_P and n_R are the number of kg-moles of gaseous matter of the products and reactants respectively.

Suppose the relationship of the internal energy U of the mixture of reactants to the mixture's temperature is the curve called 'reactants' in Fig. 20. 3 and that the second curve in the figure, called 'products' is the relationship between the internal energy of the products and their temperature. It will be seen that the vertical line 1-2 represents $(U_{PO} - U_{RO})$, in fact ΔU_0, of equation (20.4). As the reactants and products curves diverge in Fig. 20.3 it will be seen that the higher the temperature T_0 the greater will be ΔU_0.

Similarly, if the relationship of the enthalpies H of the mixture's reactants and products are as shown in Fig. 20.4, it will be seen that points 1 and there represent H_{RO} and H_{PO} and the perpendicular to the T axis represents $(H_{PO} - H_{RO})$ or ΔH_0 of equation (20.5).

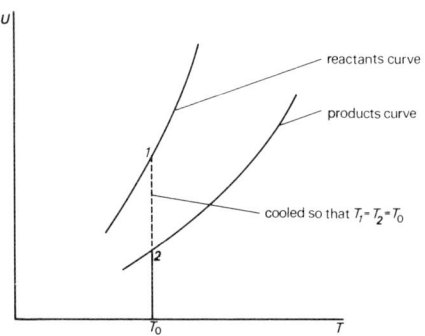

FIG 20.3   Closed system combustion

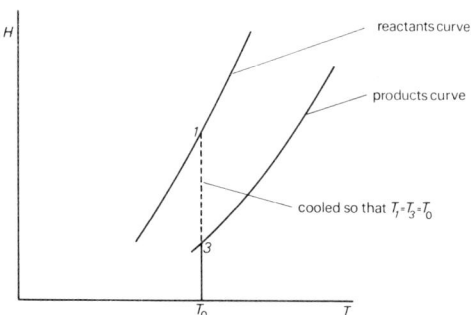

FIG 20.4   Steady-flow combustion

It is of interest to note that the slopes of the curves in Figs. 20.3 and 20.4 represent the specific internal energy per degree, c_v, and the specific enthalpy per degree, c_p, respectively, because from equations (18.4) and (18.5),

$$c_v = \frac{dU}{dT} \text{ and } c_p = \frac{dH}{dT}$$

The values of ΔU_0 and ΔH_0 determined directly or indirectly from experimental data are quoted as standard values for one kilogramme or one kg-mole of fuel at $T_0 = 25°C$.

20.6 The combustion of hydrogen with oxygen

One molecule of water has the formula H_2O which signifies that, in a molecule of water, there are two atoms of hydrogen and one atom of oxygen. If we bring together some atoms of oxygen and of hydrogen under the correct conditions water will result from the reaction (the word *hydrogen* means *water-maker*). However, it is not easy to get single atoms either of hydrogen or of oxygen because they normally exist as molecules, one molecule being a pair of atoms in these cases. That is why the symbol for a hydrogen molecule is H_2 and for an oxygen molecule is O_2. If we bring oxygen and hydrogen together under the correct conditions the gases will react and form water. Turning back to equation (20.1), F would be H_2, O would be O_2. No carbon is present to make carbon dioxide and $W = H_2O$. Equation (20.1) then becomes,

$$H_2 + O_2 \longrightarrow H_2O + energy \qquad (20.7)$$

Now this is not true because mass must be conserved and in equation (20.7) we have lost an atom of oxygen. (In fact to form the energy there is a slight decrease of mass but this is infinitesimal compared to the mass of an atom of oxygen.) The only way out of the difficulty is to supply more hydrogen molecules than oxygen molecules. For instance, if our mixture contained twice as many hydrogen molecules as oxygen molecules, equation (20.7) would read,

$$2H_2 + O_2 \longrightarrow 2H_2O + energy \qquad (20.8)$$

and no atoms have been gained or lost and mass has been conserved. Equation (20.8) states that two molecules of hydrogen react with one of oxygen to produce two molecules of water. This is a true statement of the facts so far as we know.

If we take the mass of a hydrogen atom as unit of atomic mass, then the mass of H_2 is 2 units. An atom of oxygen has been found to be about 16 units which makes its molecular mass about 32, and we can rewrite equation (20.8) as follows,

'2 × 2 units of mass of hydrogen plus 2 × 16 units of mass of oxygen combine to form 36 units of water', or if the units were kilogrammes, '4 kg of hydrogen plus 32 kg of oxygen combine to form 36 kg to water'.

20.7 The combustion of carbon with oxygen

At normal temperatures carbon exists in various forms as a solid and not as a fluid. Nevertheless carbon's reaction with oxygen is a major source of energy because carbon is abundant (coal, oil, gas, etc.) and it reacts readily with oxygen to form carbon dioxide if the combustion process proceeds to completion.

One molecule of carbon dioxide has the formula CO_2 which signifies that, in this molecule, there is one atom of carbon and two atoms of oxygen. To make carbon dioxide we must bring some atoms of carbon and of oxygen together under suitable conditions. Carbon exists as diamond, graphite and amorphous carbon; these forms are called isomers. From any of these carbon is available for reaction, although in the main graphite and amorphous carbon, combined with other elements, in fossil form (coal) is used for combustion. Carbon and oxygen atoms brought together under the correct conditions will react to form carbon monoxide CO if the oxygen is insufficient and the combustion is incomplete or to form carbon dioxide CO_2 if combustion is complete. For complete combustion, in equation (20.1), F_1 would be C, and O would be O_2, and equation (20.1) would become

$$C + O_2 \longrightarrow CO_2 + \text{energy} \qquad (20.9)$$

or

$$2C + O_2 \longrightarrow 2CO + \text{energy} \qquad (20.10)$$

for incomplete combustion, in which case per unit mass of carbon the amount of energy would be less than in the case of complete combustion by about $1 : 3 \cdot 6$. This is because of the strength of different bonds holding the atoms in the molecule. Equation (20.9) states that one atom of carbon combines with one molecule of oxygen to form one molecule of carbon dioxide and to give up some bond energy. Equation (20.10) states that two atoms of carbon combine with one molecule of oxygen to form two molecules of carbon monoxide and to give up some bond energy. These are true statements of the facts so far as we know them and they do not conflict with the idea that energy is conserved and that mass is conserved (exactly in the classical sense, nearly in an Einstein sense). An atom of carbon has been found to have about twelve times the mass of a hydrogen atom. Therefore we can rewrite equation (20.9) with regard to mass as follows:

'1 × 12 units of carbon plus 2 × 16 units of oxygen combine to form approximately $(12 + 2 \times 16) = 44$ units of carbon dioxide', or 12 kg of carbon plus 32 kg of oxygen combine to form approximately 44 kg of carbon dioxide'.

Combustion (Q and A)

Q.1. If 2 kg of carbon are burned in 16 kg of oxygen what substances are left after the combustion and what are their masses? Atomic masses of carbon and oxygen are 12 and 16.

A.1.

$$C + O_2 \longrightarrow CO_2$$

	C	O_2	CO_2	
Molecular or molar	1	1	1	moles
Mass	12	2 × 16	12 + 2 × 16 = 44	kg
	12	32	44	
Mass fractions	$\dfrac{12}{(6)}$	$\dfrac{32}{(6)}$	$\dfrac{44}{(6)}$	

The 2 kg of carbon combine with 5·34 kg of oxygen to form 7·34 kg of carbon dioxide. There are 10·66 kg of oxygen in excess of that required for combustion. So the substances after combustion will consist of 10·66 kg of oxygen and 7·34 of carbon dioxide.

Q. 2. Write down the equation for the combustion of ethylene C_2H_4 in pure oxygen. The atomic mass of hydrogen is 1.

A. 2. $$C_2H_4 + 3O_2 \longrightarrow 2CO_2 + 2H_2O$$

The number 2, 2 and 3 molecules of carbon dioxide, water and oxygen were arrived at by equating the number of carbon and hydrogen atoms and the balancing oxygen atoms respectively.

20.8 Summary

The nature of chemical reaction is discussed and the process of combustion is described both for the steady-flow and for a closed system. The effect of mixture strength and temperature of ignition have been briefly mentioned as has the energy of combustion. The combustion of hydrogen and carbon with oxygen has been studied in more detail.

20.9 Questions for the reader

Q. 1. If methane CH_4 burns in oxygen write down the equation for complete combustion.

$$[CH_4 + 2O_2 \longrightarrow CO_2 + 2H_2O]$$

Q. 2. In Question 1 if this occurred in a flow process where the flow rate of methane was 5 kg/s, what was the flow rate of the oxygen?

$$[20 \text{ kg/s}]$$

Q. 3. What was the flow rate of the products from Question 2?

$$[CO_2 = 13·75 \text{ kg/s}, H_2O = 11·25 \text{ kg/s}]$$

Q. 4. If propane C_3H_8 burns in oxygen, write down the equation for complete combustion.

$$[C_3H_8 + 5O_2 \longrightarrow 3CO_2 + 4H_2O]$$

Q. 5. If propane burns in air (molecular proportions $O_2 : N_2 :: 1 : 4$) write down the equation for complete combustion.

$$[C_3H_8 + 5O_2 + 20N_2 \longrightarrow 3CO_2 + 4H_2O + 20N_2]$$

Q. 6. Write down the equation for the complete combustion of ethyl alcohol C_2H_5OH with oxygen.

$$[C_2H_5OH + 3O_2 \longrightarrow 2CO_2 + 3H_2O]$$

Q. 7. 6 kg of benzene C_6H_6 is burnt in 25 kg of oxygen. What are the masses of the products?

$$[CO_2 = 20 \cdot 3, H_2O = 4 \cdot 2, O_2 = 6 \cdot 5 \text{ kg}]$$

Q. 8. Equal volumes of ethane C_2H_6 and acetylene C_2H_2 are burnt in air, with just enough air for complete combustion; write down the equation for this process.

$$[C_2H_6 + C_2H_2 + 6O_2 + 24N_2 \longrightarrow 4CO_2 + 4H_2O + 24N_2]$$

21

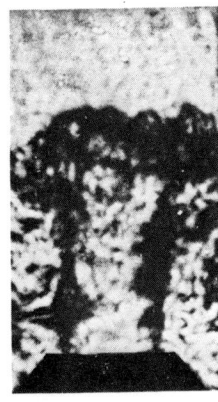

Energy from fuels

Normally engineers are interested in reactions that release energy and to this end define the term the calorific value of a fuel. So in this chapter we consider the conventional hydrocarbon-oxygen reaction and also the alternative nuclear reaction.

21.1 Calorific value

The **CALORIFIC VALUE** of a fuel may be expressed in terms of internal energy, as in equation (20.4) as ΔU_0; or in terms of enthalpy, as in equation (20.5) as ΔH_0. There are four values for the calorific value, two concerning the internal energy and two concerning the enthalpy. What follows can apply to either pair. For the change ΔU_0 there are two values, the higher calorific value, H.C.V., and the lower calorific value, L.C.V., —the differences between them are important when hydrogen forms part of the reactants.

If hydrogen exists in the fuel or in the oxidant the products of combustion will include some water vapour. For instance, if methane CH_4 were to react with oxygen the following chemical reaction would take place,

$$CH_4 + 2O_2 \longrightarrow CO_2 + 2(H_2O)_{vapour} + L.C.V. \tag{21.1}$$

This water vapour $(H_2O)_{vapour}$ will have a higher internal energy than the corresponding $(H_2O)_{liquid}$ as it contains the energy required for vaporisation, U_{lv} of water, which if the temperature were lowered below the saturation temperature (condensation temperature), then equation (21.1) would become

$$CH_4 + 2O_2 \longrightarrow CO_2 + 2(H_2O)_{liquid} + H.C.V. \tag{21.2}$$

The water vapour $(H_2O)_{vapour}$ of equation (21.1) while condensing to liquid water $(H_2O)_{liquid}$ of equation (21.2) will give up some of its internal energy, U_{lv}. Therefore the energy term L.C.V. will be lower before condensation than that H.C.V. after condensation. The former is the lower (or nett) and the latter is the higher (or gross) calorific value. These four definitions are set out in British Standard 1961, 526, and details about their experimental determination are given in British Standard 1964, 3804.

Typical values for the higher and lower calorific values are given for various fuels in Table 21.1. All the values are quoted in kJ/kg to permit direct comparison, although these may not be the units usually encountered. The values given will for the solids and liquids be ΔU_0, but for the gases be ΔH_0 because these are normally the experimental quantities measured. This as can be seen from the calculation below for methane will not matter too much as the differences between ΔU_0 and ΔH_0 are small compared with L.C.V. and H.C.V. of ΔU_0.

TABLE 21.1   Calorific values of common fuels

Substance	High calorific values (kJ/kg)	Lower calorific values (kJ/kg)
Solids		
Anthracite	29 600	28 900
Charcoal	33 700	33 100
Peat	15 900	14 500
Wood	15 800	14 300
Liquids		
Aviation fuel	47 300	44 000
Motor gasolene	48 900	43 700
Kerosene	46 500	43 400
Heavy fuel oil	42 100	39 700
Gases		
Hydrogen H_2	143 000	120 000
Methane	55 600	50 000
Ethane	51 600	47 100
Propane	50 400	46 300

In all these fuels the most important constituents are carbon and hydrogen.

Calorific values (Q and A)

Q.   The value of $(\Delta H_0)_{liquid}$ for methane is 55 600 kJ/kg of methane (see Table 21.1). Calculate the other three calorific values for methane.
A.   The equation for the combustion of methane in oxygen is

$$CH_4 + 2O_2 \longrightarrow CO_2 + 2H_2O$$

So for $(12 + 4)$ kg of methane, $2(2 + 16)$ kg of water are formed. From tables h_{lv} for water at 25°C is $2\,442$ kJ/kg.

$$(\Delta H_0)_{vapour} = 55\,600 - \frac{36}{16} \times 2\,442 \text{ kJ/kg of methane}$$

$$= 50\,100 \text{ kJ/kg (cf. experimental values, Table 21.1)}$$

(*Note:* The difference between these is about 10 per cent.)

Remembering the relationship of equation (18.6), which can be written in the form

(a)
$$\Delta H_0 = \Delta U_0 + R_0 T_0 \, (n_p - n_R)$$

then
$$(\Delta U_0)_{liquid} = 55\,600 - 8 \cdot 314 \times 298 \times (1-3)/16 \text{ kJ/kg}$$

$$= 55\,910 \text{ kJ/kg}$$

(b) and
$$(\Delta U_0)_{vapour} = 50\,100 - 8 \cdot 314 \times 298 \times (3-3)/16 \text{ kJ/kg}$$

$$= 50\,100 \text{ kJ/kg (no change)}$$

(*Note:* The value of n_R changes as the water is condensed in (a) and $(\Delta H_0)_{vapour} = (\Delta U_0)_{liquid}$ in (b) because the number of moles of reactant and product are equal. Compare this with Question 7 at the end of the chapter.)

21.2 Stoichiometry

Stoichiometry is the part of chemistry dealing with the compositions of substances and more particularly with the balancing of chemical equations. The level of dexterity required by engineers is elementary but necessary to interpret analyses of combustion processes.

The best method of discussing this technique is to consider a number of examples. Consider the combustion of a solid fuel that is mainly carbon in one form or another. If one molecule of carbon (one atom) is burned it will combine with one molecule of oxygen to produce one molecule of carbon dioxide and release some energy—see equation (20.9). However, if one has twice as many carbon molecules as oxygen molecules the product will be two molecules of carbon monoxide—see equation (20.10).

$$C + O_2 \longrightarrow CO_2 + \text{energy} \qquad (20.9)$$

$$2C + O_2 \longrightarrow 2CO + \text{energy} \qquad (20.10)$$

If the ratio of carbon to oxygen molecules is between 1 and 2 then the product will be a mixture of carbon dioxide and carbon monoxide:

$$aC + bO_2 \longrightarrow xCO_2 + yCO \qquad (21.3)$$

If one wishes to determine values of x and y, knowing a and b or vice versa, this is possible by applying the principle of conservation of mass, which gives

$$a = x + y \text{ by balancing the carbon molecules}$$

and $$b = x + \frac{y}{2} \text{ by balancing the oxygen molecules}$$

This technique is known as using a carbon and an oxygen balance.

When equation (20.9) or something like it describes the reaction, where precisely the right amount of oxygen is present to combine with the fuel, to give complete combustion of the fuel—no excess oxygen nor excess fuel—the mixture is said to be a **Stoichiometric mixture**, although this use of the word is more restricted than its full meaning.

21.3 Dissociation

When carbon is burning in oxygen the process that actually occurs is most complicated. A very simple version might be considered by supposing that the reaction occurs in two stages. In the first stage the carbon combines with oxygen to form carbon monoxide and, subsequently, carbon monoxide combines with more oxygen to form carbon dioxide:

$$C + O_2 \longrightarrow CO + O + \text{energy} \longrightarrow CO_2 + \text{more energy} \qquad (21.4)$$

Both these stages are exothermic and energy is released. However, simultaneously to an extent depending on the temperature at the time, the carbon dioxide is dividing to form carbon monoxide and oxygen again:

$$CO + O - \text{energy} \longleftarrow CO_2 \qquad (21.5)$$

which, as expressed by negative energy in the equation, is an endothermic reaction because it absorbs energy. At any temperature the proportions of CO_2, CO and O_2 adjust themselves to an **equilibrium** so that the rate of the forward reaction as in equation (21.4) equals the rate of the backward reaction as in equation (21.5). In this equilibrium situation there is apparently no combustion occurring because the number of CO_2 molecules being formed is equal to the number dissociating:

$$CO + O \longleftrightarrow CO_2 \qquad (21.6)$$

and the net result is neither exothermic nor endothermic because there is no apparent energy change. No work is done and it is adiabatic. In a flame, which is not adiabatic and not in equilibrium, the number of CO_2 molecules being formed is greater than the number dissociating. However, because dissociation does occur the flame is at a lower temperature than it would be if dissociation did not occur. The higher the temperature of the system

$$CO + O \longleftrightarrow CO_2 \qquad (21.7)$$

$$H_2 + O \longleftrightarrow H_2O \qquad (21.8)$$

the more likely are you to find greater quantities of CO and O from equation (21. 7), and H_2 + O from equation (21. 8) (CO, O, H_2 are not necessarily the actual species present, they might be O_2 or OH radicals). The temperatures of chemical equilibrium during combustion, in the sense described above, are capable of prediction but such calculations are beyond the capacity of this book.

21. 4 Nuclear reactions

An atom is a particle that defines the nature of an element (a particular substance), and is thought to be comprised of many elementary particles. The most important of these are called electrons, protons and neutrons. These elementary particles form two groups in the atom, a concentrated group called a nucleus at the centre of the atom and another scattered group that moves around the nucleus at some distance from it. If the nucleus is one of relatively large mass, such as the nucleus of an atom of uranium, it is found that the atom is so large that it is unstable. Imagine a classroom packed with restless children into which another child is squeezed, this act of being enough to cause a riot. It is the same with the nucleus of a large atom into which another elementary particle is squeezed, this has the effect of breaking down the large atom and releasing a large quantity of energy. This process of splitting large unstable atoms with elementary particles causing energy to be released is called fission. In a nuclear reactor using uranium fuel the amount of useful energy liberated by 1 g of uranium fuel mixture corresponds to about 10 kg of coal being burnt.

Another type of nuclear process is the type that goes on to give tne massive quantities of energy that the sun liberates and is called fusion. This occurs when light atoms, for example hydrogen or helium, are brought into contact.

In both fission and fusion the process will involve a decrease of mass, and the quantitative relationship between energy E released and the reduction of mass m is:

$$E = mc^2 \tag{21.9}$$

where c is the velocity of electromagnetic waves. The general statement can be made that circumstances can be found in which heavy atomic nuclei can be made to divide and cause a release of energy, and others (not yet attained by man) in which light atomic nuclei can be made to associate also causing a release of energy by a loss of mass. Nuclear reactions are the only situations where for all practical purposes conservation of mass is not maintained in a thermodynamic process.

21. 5 Summary

Calorific values of fuels have been defined and the differences between the higher and lower values explained. In Table 21. 1 a comparison of calorific values of some common fuels is given. Stoichiometry has been referred to and the restricted use of the term stoichiometric mixture has been defined. Dissociation has been explained and the fact that, because it is occurring simultaneously with combustion in a flame, the temperature of a flame is thereby made cooler. Finally nuclear fission and fusion were briefly described.

21.6 Questions for the reader

Q.1. What type of atomic nuclei are likely to produce nuclear energy by

(a) Fusion
(b) Fission
(c) Are unlikely to produce energy?

[light, heavy and medium atoms]

Q.2. 4 kg of carbon are burnt with 10 kg of oxygen. If there is no unburnt carbon or oxygen as product, what are the products?

[$CO_2 = 12 \cdot 8$, $CO = 1 \cdot 2$ kg]

Q.3. Propane C_3H_8 is burnt in air. The products of combustion as a molar analysis are:

CO_2	CO	H_2O	N_2
1	2	4	16

What are the molar proportions of the reactants?

[$C_3H_8 : O_2 : N_2 :: 1 : 4 : 16$]

Q.4. For Question 3 what is the mass ratio of propane to air?

[$1 : 13 \cdot 1$]

Q.5. For a stoichiometric mixture of propane and air what would be the fuel : air ratio?

[$1 : 16 \cdot 4$]

Q.6 Using the value of the lower calorific value for hydrogen H_2 given in Table 21.1, calculate the corresponding higher calorific value.

[142 000 kJ/kg]

Q.7. Using the value of the lower calorific value given for propane C_3H_8 in Table 21.1, calculate the other three calorific values.

[$(\Delta H_0)_{vapour} = 46\,300$ (given), $(\Delta H_0)_{liquid} = 50\,300$
$(\Delta U_0)_{vapour} = 46\,244$, $(\Delta U_0)_{liquid} = 50\,469$ kJ/kg]

Q.8. Using the lower calorific value of methane given in Table 21.1, calculate the adiabatic flame temperature of methane burning in a stoichiometric mixture with air if the reactants are at 25°C, using

c_p for $CO_2 = 1 \cdot 2$ kJ/kg K
c_p for $H_2O = 2 \cdot 4$ kJ/kg K
c_p for $N_2 \quad = 1 \cdot 2$ kJ/kg K

[2 260 K compared with a more accurately predicted value of 2 130 K]

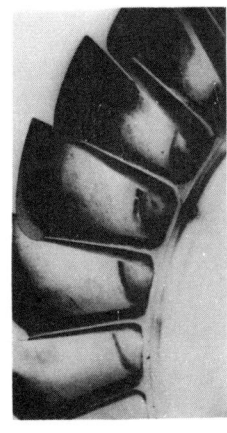

22 *A gas power cycle*

A heat engine working as a closed-cycle power plant is analysed and compared with an open-cycle gas power plant. Temperature-entropy diagrams are examined for the reversible and irreversible cases. Arising from this, isentropic efficiencies are defined for turbines and compressors.

22.1 A closed-cycle gas-turbine power plant

A closed-cycle gas-turbine power plant is one form of heat engine. A heat engine (Fig. 22.1) is a system in which there is a working fluid undergoing a cyclic process and across the boundary of which heat energy q_1 enters the system, and heat energy q_2 leaves the system and a transfer of work energy w out of the system takes place.

In Fig. 6.3, and again in Fig. 22.2, a power cycle is shown that consists of a working fluid—gas in the case of a gas power cycle—circulating through the

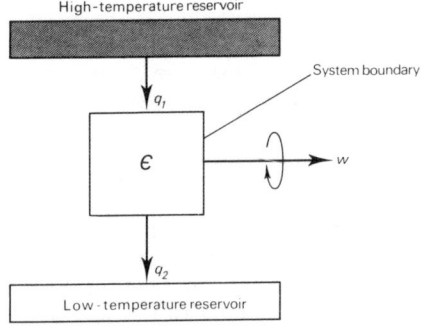

FIG 22.1   The heat engine

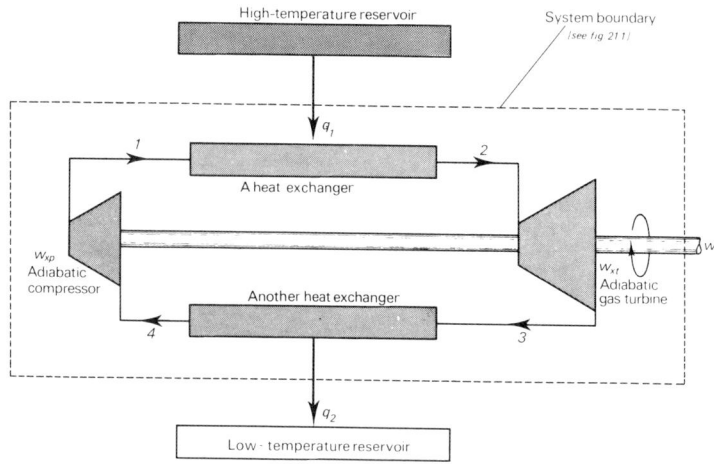

FIG 22.2   The components in a gas power plant

following series of components,

(a)   A heat exchanger
(b)   An adiabatic expander
(c)   Another heat exchanger
(d)   An adiabatic compressor.

These components, when used in a gas power plant, are shown in Fig. 11.2 and again in Fig. 22.3. The components are the following,

(a)   A heater in which the gas as it passes through the chamber takes in q_1 units of heat energy per unit mass of working fluid

(b)   A gas turbine in which the gas exchanges no heat energy with its surroundings (for this reason the process is called adiabatic), but does w_{xt} units of work energy per unit mass of working fluid

(c)   A cooler in which the gas gives out q_2 units of heat energy per unit mass of the working fluid

(d)   A compressor in which the gases undergo an adiabatic process during which w_{xp} units of work energy per unit mass of working fluid is done on it.

Figure 22.3 shows the gas power plant of Fig. 22.2 in more detail. In it the working fluid, a gas, moves around the closed cycle 1-2-3-4-1. Energy is supplied by the combustion of, say, a hydrocarbon fuel and air in a combustion chamber. The products of combustion at a high temperature enter the heat exchanger and pass across its tubes to exhaust. In passing through the heat exchanger these products transfer energy by heat to the working fluid. The temperature of the working fluid is thereby increased as it passes from 1 to 2 through

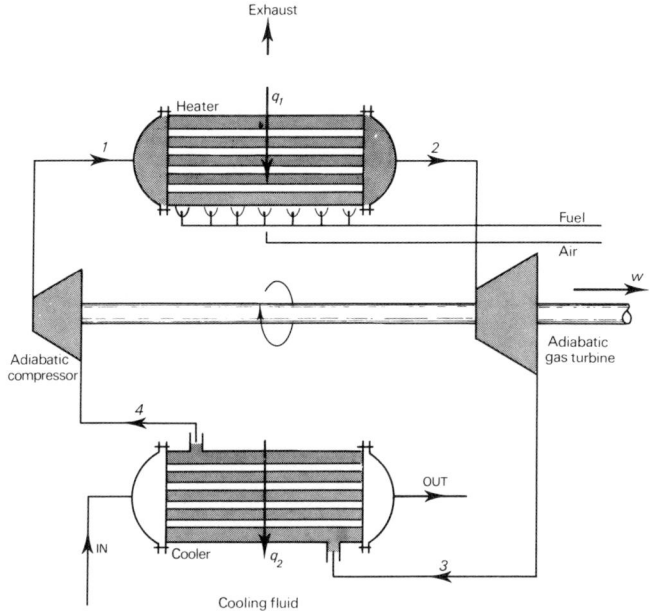

FIG 22.3   A closed-cycle gas power plant

the tubes of the heat exchanger. From 2 the gas enters the adiabatic gas turbine, doing work as it moves to point 3 where it passes across the outside of the tubes of a cooler. A coolant, which might be water or air, passes through the tubes and takes in energy transferred from the gas. The coolant's temperature increases as it passes through the cooler and the temperature of the gas is reduced. A compressor takes in the gas from point 4 and compresses it adiabatically to its original state at point 1.

22.2 Energy transfers in a gas power cycle

If we now examine more closely the energy exchanges between the circulating gas and its surroundings we find it easy enough to imagine where the energy w goes to. It is used for propulsion in the case of aircraft, ships and automobiles. w is the sum of the two components w_{xt} and w_{xp}; w_{xt} is the work done by the gas on the blades of the turbine, and w_{xp} is that done on the gas by the compressor. In fact,

$$w = w_{xt} + w_{xp} \tag{22.1}$$

The numerical value of w_{xt} is positive and of w_{xp} is negative so that w, the nett work done per unit mass of the working fluid, is always smaller than the greatest positive work done.

There is a ratio called the **Work ratio**, r_w, which is kept as near to 1 as possible by keeping at a minimum the amount of energy that must be fed back into the working fluid via the compressor. The value of r_w is given by,

$$r_w = \frac{w_{xt} + w_{xp}}{w_{xt}} \tag{22.2}$$

The examination of the components in a cycle is a study of a steadily flowing gas passing through a number of control volumes each of which encloses a component of the power plant. Therefore, for each control volume, the steady-flow energy equation (8.6) applies,

$$Q - W_x = \Delta H + \Delta K + \Delta Z \tag{22.3}$$

or, per unit time,

$$\dot{Q} - \dot{W}_x = \Delta \dot{H} + \Delta \dot{K} + \Delta \dot{Z}$$

or, from this and equations (8.16), (3.4) and (3.5),

$$\dot{Q} - \dot{W}_x = \dot{M} c_p (T_2 - T_1) + \dot{M} \left(\frac{V_2^2 - V_1^2}{2} \right) + \dot{M} g (z_2 - z_1) \tag{22.4}$$

where $\dot{M}$ is the rate of flow of working fluid through the control volume. Sometime but not always $\Delta \dot{K}$ and $\Delta \dot{Z}$ are small when compared with $\Delta \dot{H}$ and one or both may be neglected, just as other terms such as the surface tension and potential energies due to electric and magnetic fields have previously been neglected, making equation (22.4) into

$$\dot{Q} - \dot{W}_x = \dot{M} c_p (T_2 - T_1) \tag{22.5}$$

or, per unit mass of working fluid per unit time,

$$\dot{q} - \dot{w}_x = c_p (T_2 - T_1) \tag{22.6}$$

Components (Q and A)

Q.1. The heater between points 1 and 2 of the cycle in Fig. 22.3 has air, its working fluid, passing through the heat exchanger tubes at the rate of 2 kg/s. This air comes from the compressor at 20°C with negligible velocity, and is heated by the walls of the heat exchanger tubes as it passes through the tubes. Its temperature rises to 550°C. Across the outside surfaces of the tubes 4 kg/s of the products of combustion from a gas fired furnace strike the outer surfaces of the first tubes at

860°C and leaves the surface of the last tubes at 600°C. The velocities and there-fore the kinetic energies of the products of combustion are negligible. Draw up an energy balance. Take $c_p = 1 \cdot 060$ kJ/kg K for air and $1 \cdot 101$ kJ/kg K for the products of combustion.

If the energy balance shows an apparent loss of energy perhaps you have been wrong in considering ΔK negligible. If you come to the conclusion that K_1 is negligible but K_2 is not, find the velocity of the working fluid at point 2.

A. 1. For the working fluid, air, the rate of increase of enthalpy = $2 \times 1 \cdot 060$ $(550-20) = 1\,102$ kJ/s (or kW) and the rate of loss of enthalpy by the products of combustion = $4 \times 1 \cdot 101$ $(860-600) = 1\,144$ kW. The rate of loss of enthalpy by the products of combustion that is not accounted for by an increase of enthalpy by the working fluid $= 1144 - 1123 = 21$ kW.

Assuming that the loss of energy by heat to the surroundings is negligible the energy lost has probably gone into kinetic energy in the working fluid. It is stated in the question that V_1 is negligible

$$2\left(\frac{V_2^2 - O}{2}\right) = 21 \times 10^3 \text{ J/s (or W)}$$

$$V_2 = 145 \text{ m/s}$$

Q. 2. In the same plant as in Question 1 the air leaves the turbine at 400°C and at a velocity of 100 m/s. What is the power output $\dot{W}_{xt}$ of the turbine alone? Changes of gravitational energy can be neglected.

A. 2. The power $\dot{W}_{xt}$ of the turbine can be found from equation (21. 4), taking,

$$\dot{Q} = O \text{ because the turbine is adiabatic}$$

$$\dot{W}_x = \dot{W}_{xt}$$

$$\Delta \dot{Z} = O \text{ because changes in Z are negligible}$$

$$\Delta \dot{H} = \dot{M} c_p \Delta T = 2 \times 1 \cdot 060 \ (400-550) = -318 \text{ kW}$$

$$\Delta \dot{K} = \frac{2 \ (100^2 - 145^2)}{2 \times 10^3} = -11 \text{ kW}$$

Therefore from equation (21. 4)

$$O - \dot{W}_{xt} = -318 - 11$$

$$\dot{W}_{xt} = 329 \text{ kW}$$

22. 3 Temperature-entropy diagram

In the four components mentioned in this chapter the working fluid of a gas turbine undergoes four idealised simple energy-transfer processes. These are:

(a) A reversible simple heat transfer process in the heater, 1-2
(b) A reversible simple work transfer (isentropic) in the turbine, 2-3

(c) A reversible simple heat transfer process in the cooler, 3-4
(d) A reversible simple work transfer process (isentropic) in the compressor, 4-1.

The processes, which cause a change of state in the working fluid, are shown on a temperature-entropy diagram in Fig. 22. 4. In practice a reversible isothermal heat transfer process is impossible to obtain. Remember that the heat transfer process can only be reversible if it takes place along a zero temperature gradient, which, if the high-temperature reservoir were at constant temperature, would be represented by the line 1″-2 in Fig. 22. 4. In practice, however, it is not the temperature but the pressure that remains constant throughout the heater and cooler. Consequently the changes 1-2 and 3-4 more nearly represent the state paths of the gas passing through the heater and through the cooler than do the lines 1″-2 and 3″-4. So the more realistic cycle is 1-2-3-4 shown by dashes in Fig. 22. than the ideal cycle 1″-2-3″-4.

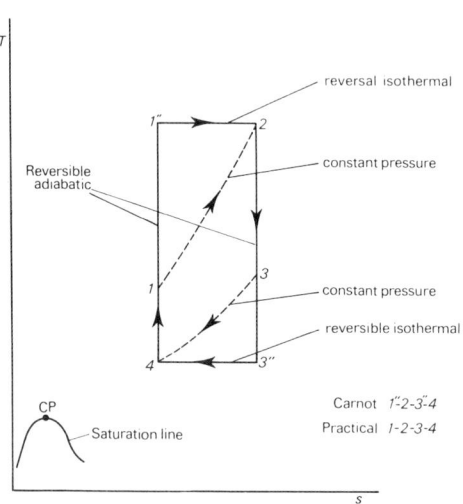

FIG 22. 4   The state path of the gas in a gas power plant. The scale on this figure has been distorted a little for clarity. The line scale more like that of Figure 24. 3.

In Fig. 22. 4 it can be seen that the state of the working fluid as it goes around the cycle is always far from the saturation lines and far from the critical point. The fluid can therefore be treated as a gas and assumed to obey the ideal gas rule (see Chapter 18).

22. 4 Closed and open cycles

The cycle described so far in this chapter has been what is called
A closed cycle. The same fluid goes around the plant from 1 to 2, 3, 4 and back to

again. A secondary fluid such as the products of combustion must be used, as in Fig. 22.3, to supply energy to the working fluid. This energy is supplied by heat transfer in the heater and for this to occur the surface area of the tubes in the heater would have to be sufficiently large for the heat transfer to take place. For instance, if the heat transfer coefficient (that is the energy transferred per unit area per degree of difference in temperatures of the fluids) were as high as 30 W/m²°C, the surface area of the tubes for a power as little as 100 kW and a temperature difference of 100°C would be 33 m²—a formidable size. Instead of the heater shown in Fig. 22.3 imagine the arrangement shown in Fig. 22.5. Here the combustion chamber and heater of Fig. 22.3 have been combined so that fuel is injected into the working fluid and takes the oxygen required for combustion from the working fluid which then becomes the oxidant.

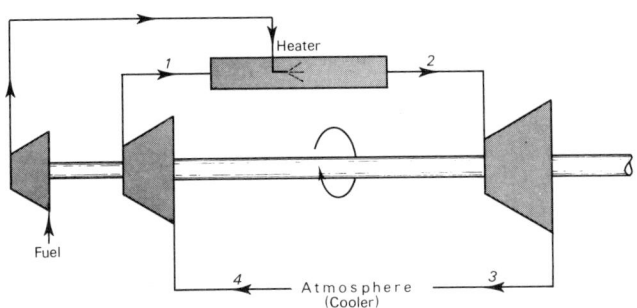

FIG 22.5   An open-cycle gas power plant

One consequence of using the working fluid as the oxidant for combustion of the fuel is that the fluid leaving the heater at 2 is not the same as the fluid entering it at 1. The working fluid at 2 contains less oxygen than when it was at 1 and owing to the combustion process it contains various products of combustion at 2. Also, because the fuel is added in the combustor, the mass flow-rate will vary between 1 and 2. If the cycle were closed like the cycle of Fig. 22.3 the gas entering the heater at 1 would have a diminishing quantity of oxygen and an increasing quantity of the products of combustion. Clearly such a continuous change cannot be permitted to continue for long, because finally there would be insufficient oxygen in the gas to burn the fuel and the plant would stop working. In practice when air is the working fluid the cooler is replaced by the Earth's atmosphere. The products of combustion are discharged from point 3 of Fig. 22.5 into the atmosphere and fresh air is taken in from the atmosphere at 4 where it enters a compressor. At first sight it may appear that the cycle is broken but in fact one can imagine the whole of the Earth's atmosphere as a gigantic component that not only acts as a cooler but also as a reconstitutor for the fuel and the air. The state of the gas in an open-cycle gas power plant, looked at in this way, follows a cyclic path very similar to 1-2-3-4-1 on the temperature-entropy diagram of Fig. 22.4 as does the gas in a closed cycle gas power plant. The energy transfer and consequent changes of enthalpy, kinetic and gravitational energies are still carried out in accordance with equation (22.3).

22.5 Isentropic efficiency

The change of state from 2 to 3 shown in Fig. 22.4 and repeated in Fig. 22.6 is the state path of the working fluid in a gas power plant as it passes through a reversible adiabatic turbine. This, however, is an ideal case in that the turbine is considered adiabatic and reversible. This is not attainable in practice. The actual state path of the gas as it passes through the turbine is not 2-3 in Fig. 22.6 but 2-3'. The gas starts and finishes at the same pressure in the reversible process 2-3 and in the irreversible process 2-3'. The processes undergone are both adiabatic and in the ideal case 2-3 it is reversible too and therefore isentropic (see equation (14.3)), but in the irreversible case the entropy increases (see equation (14.5)) as for the process between 2 and 3'. The final pressure at 3' is the same as that at 3 but the entropy has increased.

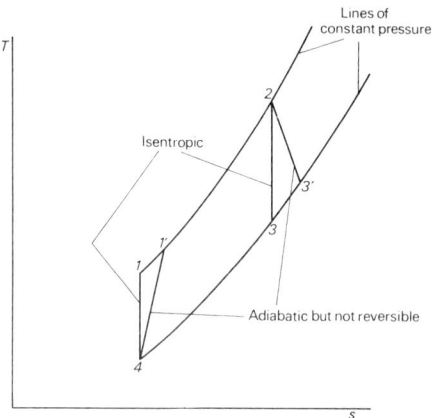

FIG 22.6   A more realistic state path

Similarly during the compression process the ideal, reversible and adiabatic, process is shown by the line 4-1, but the irreversible adiabatic process is shown by 4-1'. During the process 4-1' the entropy has increased, while during the ideal process 4-1 the change of state is isentropic. In both cases the change in pressure is the same and just as 3 and 3' lie on the same pressure line so do 1 and 1'.

If we consider the expansion from 2 to 3 we find it adiabatic so $Q = 0$ and if changes in kinetic and gravitational energies are negligible we can write equation (22.3)

$$W_{xrt} = -\Delta h$$

using W_{xrt} instead of W_x to remind us that the change is reversible and refers to a process in the turbine. In units of energy per unit mass this equation becomes,

$$w_{xrt} = -\Delta h$$
$$= -(h_3 - h_2) \tag{22.7}$$

For the irreversible expansion from 2-3′ equation (22.7) becomes

$$w_{xt} = -(h_{3'} - h_2) \tag{22.8}$$

The expression (22.7) gives the work done in the course of an isentropic—that is reversible and adiabatic—process, and (22.8) is the work done in the course of an irreversible adiabatic but more practical process. The ratio

$$\frac{h_{3'} - h_2}{h_3 - h_2} = \frac{w_{xt}}{w_{xrt}} = \eta_I \tag{22.9}$$

and this ratio η_I is called the **Isentropic efficiency**, although it is not the efficiency within the definition given by equation (6.7). It is the ratio of the work energy given out during an actual adiabatic process to the maximum work energy that could possibly be given out.

Similarly for the compressor, the work done isentropically is

$$w_{xrp} = -(h_1 - h_4)$$

and that done in practice is

$$w_{xp} = -(h_{1'} - h_4)$$

In this case the isentropic efficiency for a compressor is given by

$$\frac{h_1 - h_4}{h_{1'} - h_4} = \frac{w_{xrp}}{w_{xp}} = \eta_I \tag{22.10}$$

It should be noted from the above that we have followed the unfortunate custom of using two meanings for η_I. This is because w_x is greatest when w_{xt} is a maximum and w_{xp} is a minimum. Hence the process is most efficient when w_{xt} is maximum and w_{xp} is a minimum. Therefore w_{xt} occurs in the numerator and w_{xp} in the denominator for η_I. The designer's aim is to maximise the efficiency of the cycle, η.

The isentropic efficiency of a turbine is

$$\eta_{\rm I} = \frac{\text{work actually done } by \text{ gas adiabatically}}{\text{work that would have been done } by \text{ gas isentropically}}$$

and for the compressor

$$\eta_{\rm I} = \frac{\text{work that would have been done } on \text{ gas isentropically}}{\text{work actually done } on \text{ gas adiabatically}}$$

22.6 Thermodynamic cycle of a reciprocating engine

At first sight it appears that the reciprocating engine—petrol, oil or steam—could not, because of the intermittent opening and closing of its valves, be concerned with steady flow. In fact many analyses of the behaviour of such pieces of equipment quite correctly consider a system of a gaseous (or vapours and gase mixture in the engine's cylinder. This is quite correct when applied to a fixed ma of gas, for instance after the inlet valve closes and before the outlet valve opens. During that time no mass crosses the boundary of the system (see Chapter 23). T such a system the equation (3.2) applies

$$Q - W = \Delta U + \Delta K + \Delta Z \qquad\qquad (22.1)$$

Indeed a treatment of one cycle of such an engine could be based on this equation. Such a treatment is applied in the next chapter. However, for the use of reciproca ing engines as components in thermodynamic cycles an approximation is made. T speed of modern engines working at their most efficient and economic is such that the steady-flow energy equation, equation (8.6), represents the approximate truth; also multi-cylinder engines improve the overall conditions by producing more nearly steady conditions. If the reciprocating engine were enclosed in a control volume and the engine operated at high speed, the fuel and air would effectively flc steadily into and the products of combustion steadily out of the control volume because the periodic events are occurring so rapidly. Many of the series of processes depicted by Fig. 6.3 would then take place within the control volume, as shown in Fig. 22.7. The control volume shown in Fig. 22.7 would enclose within itself basically a single component in which a successive series of processes occu The whole cycle of processes might be,

(a) First process—Air from the cooler, which might be the atmosphere, is drawn into a chamber in which it is mixed with fuel.
(b) Second process—The mixture of fuel and air then enters the cylinder where it is compressed by the piston from state 4 to state 1
(c) Third process—The compressed mixture remains in the cylinder at point 1, but the cylinder changes roles from that of compressor to that of heater or combustion chamber. To cause this change of process, a sparking plug in the cylinder head is operated so initiating

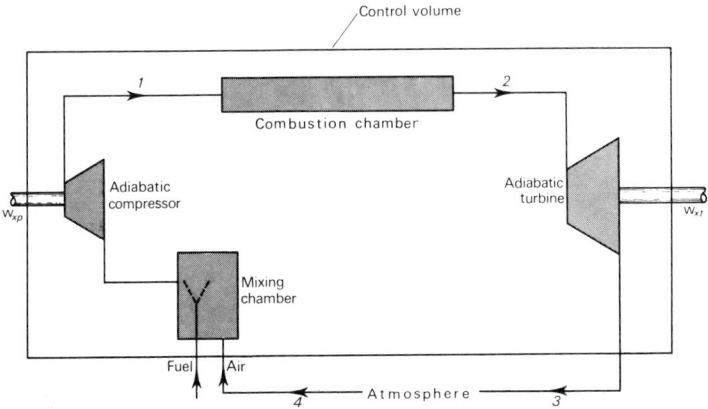

FIG 22.7   Four roles of a reciprocating engine

the combustion process which occurs quite rapidly with the state of
the mixture changing from 1-2, both temperature and pressure rising
rapidly. The burning may be considered to occur so quickly that this
process occurs at constant volume.

(d) Fourth process—The role of the cylinder changes again to that of
expander as the hot gas pushes the piston out doing work against a
restraining force hence changing its own state from 2 to 3.

At the end of the fourth process the products of combustion leave the engine via an
exhaust pipe to the atmosphere which is acting as a cooler as in the open cycle of
Fig. 22.5. Thus it is seen that three of the four processes mentioned above as
being undergone in a closed system are the same processes of compressing, heating
and expanding that occurred in the open-cycle gas-turbine plant. In this case three
of the processes—compressing, heating and expanding—take place inside one component. For the working fluid passing through this multi-purpose component plus
the mixing chamber equation (22.3) applies in the form

$$\dot{Q} - \dot{W}_X = \dot{M}(h_3 - h_4) + \dot{M}\left(\frac{V_3^2 - V_4^2}{2}\right) + \dot{M}g(\bar{z}_3 - \bar{z}_4) \qquad (22.12)$$

22.7 Summary

The heat engine called a gas power cycle and its components have
been described. The steady-flow energy equation is applied to each component and
the cycle undergone by the gas shown on a temperature-entropy diagram. Closed
and open cycles are discussed and isentropic efficiencies defined. The position of
a reciprocating engine in a thermodynamic cycle is introduced.

22.8 Questions for the reader

Q.1. In an open-cycle gas power plant the turbine does 300 kJ of work per unit mass of the working fluid. If the work ratio is 0·40 what work is done in the compressor? Assume the air enters at a pressure of one bar.

[180 kJ/kg]

Q.2. In Question 1 the air is drawn into the adiabatic compressor at a temperatur of −10°C. What is its temperature on leaving the compressor? The specific enthalpy per degree of the air is 1·02 kJ/kg K.

[166°C]

Q.3. What is the isentropic efficiency of the compressor of Questions 1 and 2 if the pressure of the air leaving it is 3 × 10⁵ N/m²? $c_v = 0·720$ kJ/kg K.

[0·56]

Q.4. Draw lines on a T-s diagram for the reversible and irreversible compressors described in Questions 1 to 3.

[See Fig. 22.8]

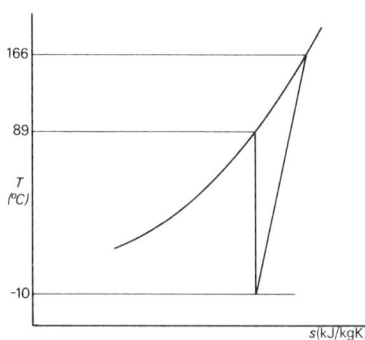

FIG 22.8   Answer to question 4

Q.5. A Carnot cycle works between two energy reservoirs at temperatures of 527°C and 27°C with a work ratio of 0·75. If $\dot{Q}_1$ for the cycle is 80 kW, what are the values of $\dot{W}_T$ and $\dot{W}_P$?

[66·7, − 16·7 kW]

Q.6. If operating between the same energy reservoirs as those of Question 5 there were an engine in which the isentropic efficiencies of the feed pump and turbine were 0·80, with the same Q_1, what would the work ratio of that plant be?

[0·61]

Q. 7. With regard to the changes in value of the overall efficiency η of a cycle as the isentropic efficiencies η_I of the turbine and compressor change,

> (a) Consider a steam turbine power plant where $\eta = 0\cdot40$, η_I for the turbine and compressor being $0\cdot80$. How does η change when both η_I drop first to $0\cdot70$ and then to $0\cdot60$ (remember that for a steam turbine $w_p \simeq 0$).
>
> (b) Consider a gas turbine power plant where $\eta = 0\cdot40$, η_I for the turbine and compressor being $0\cdot80$, when η_I drops to $0\cdot70$ and $0\cdot60$ (assume a constant work ratio $r_w = 0\cdot35$). How does η change?
>
> [η_{steam} drops to $0\cdot35$ then to $0\cdot30$, and
> η_{gas} drops to $0\cdot35$ and then to $0\cdot30$]

23 Reciprocating engines

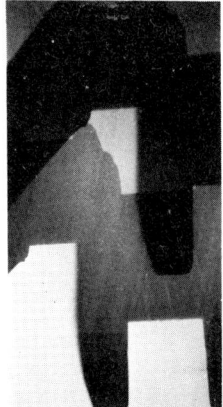

Having considered the gas as a working fluid in a turbine power plant we now consider the gas as the working fluid in a reciprocator. A practical reciprocator is described and then compared with the theoretical models, the Otto and Joule cycles.

23.1 The cylinder and piston

The cylinder A of a reciprocating engine is shown in Fig. 23.1. A piston P, also cylindrical, is fitted into the cylinder and moves from the extreme position, (a), shown in Fig. 23.1 and called **Inner dead centre**, through position (b) to the other extreme position shown in Fig. 23.1 (c) called **Outer dead centre**, and back again through (d) to inner dead centre (a). The volume of the cylinder above the piston between positions (a) and (c), that is ($V_c - V_a$) is called the **Swept volume.** A schematic side elevation of the cylinder is shown in Fig. 23.2 so that the connecting

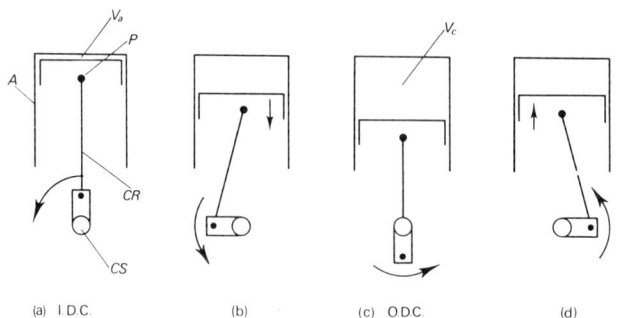

(a) I.D.C.   (b)   (c) O.D.C.   (d)

FIG 23.1   A reciprocating engine

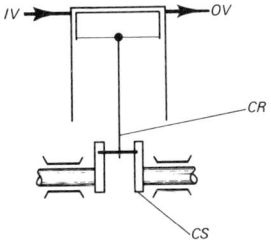

FIG 23.2   Another view of a reciprocating engine

rod CR and the crankshaft CS can be seen. The working fluid expands in the cylinder driving the piston in an outward direction from (a) to (c) via (b). The pressure of the working fluid on the top of the piston causes the piston to exert a force on the connecting rod which in turn drives the crank shaft CS. For admitting the working fluid there is an inlet valve IV, and for allowing the working fluid to be removed from the cylinder there is an outlet valve OV.

An important parameter is the compression ratio because on this parameter depends the effectiveness of the combustion, and it also affects the maximum pressure and the maximum temperature of the working fluid in the cylinder. The **Compression ratio** is defined as

$$\frac{\text{the clearance volume at I.D.C.}}{\text{the swept volume plus the clearance volume}}$$

or in terms of Fig. 23.1, the compression ratio $= V_a/V_c$.

23.2 The petrol engine

The four-stroke petrol engine may be said to begin with the piston at inner dead centre shown as (a) in Fig. 23.1. In this position one may assume that both the inlet and outlet valves IV and OV are closed, and the volume V_a of the cylinder above the piston is very small and contains residual gas from the previous cycle at a low pressure because discharge to the atmosphere has just finished. This point is represented by a_0 on the pressure-volume diagram in Fig. 23.3. Because of kinetic energy stored in the flywheel and the flywheel's connection to the piston via the crankshaft and connecting rod, the piston moves from position (a) and the inlet valves IV opens. This movement causes a new charge of fuel and air to enter the cylinder, and the maximum charge occurs when the piston reaches position (c). This is shown in Fig. 23.3 as the piston moves from a_0 to c via b_0. When position (c) is reached the inlet valve closes. At position (c) the piston is at outer dead centre and the piston then moves upwards again with both the inlet and outlet valves closed. This means that the flow of mixture has stopped and a fixed mass of gas is being considered which should thermodynamically be regarded as a system. The piston will move from (c) to (a) via (d) and the state of the gas changes from c to a_1 via d. Just before the piston arrives at (a) a sparking device

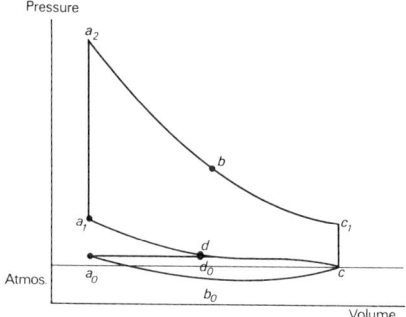

FIG 23.3   The pressure-volume diagram of a four-stroke petrol engine

in the cylinder ignites the mixture causing a relatively large increase in temperature and therefore of pressure of the gas whose state point rises to a_2 in Fig. 23.3. This occurs in so short a period of time that the piston can be said not to have moved significantly but to have remained in position (a). The high pressure of the gas now drives the piston downward from position (a) in Fig. 23.1 to (c) via (b), the products of combustion going from state a_2 to b and on to c_1. With the piston again in position (c) the outlet valve opens and the pressure in the cylinder drops quickly to c, completing the system's cycle c, d, a_1, a_2, b, c_1, c. The piston changes direction at (c) and starts to move inward again driving the combustion products out of the cylinder through the open exhaust valve. As the piston again approaches position (a) the outlet valve closes leaving only residual burnt gas in the cylinder. The cycle undergone by the system c, d, a_1, a_2, b, c_1, c is the cycle that we are considering in this chapter.

23.3 The Otto cycle

The cycle shown in Fig. 23.3 is approximately the same as the arbitrarily drawn cycle shown in Fig. 23.4 and known as the Otto cycle. State points a_1, a_2, c_1 and c of the practical cycle correspond to similar points in the idealised cycle of Otto. In the ideal cycle Otto imagined that there was a fixed mass of a perfect gas and this undergoes the following four processes shown in Fig. 23.4:

c-a_1 Isentropic compression from v_1 to v_2.
a_1-a_2 Addition of energy by heat at constant volume causing an increase of temperature and pressure. This is considered at constant volume because it compares with the practical case in which combustion occurs very rapidly.
a_2-c_1 Isentropic expansion from v_2 to v_1.
c_1-c   Energy is withdrawn from the system at constant volume while the temperature and pressure fall.

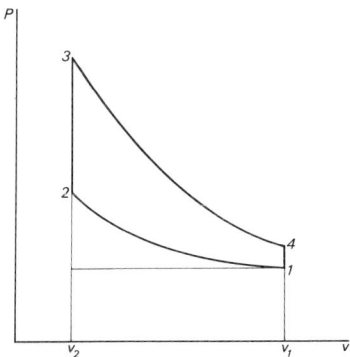

FIG 23.4   The Otto cycle

If this idealised case is compared with the practical case the main difference is that the last process is performed by getting rid of the exhaust products and taking in a new air/fuel charge because, in the step a_1-a_2, energy is released internally by combustion making the old charge useless for the next cycle.

The efficiency of a practical petrol engine is below that of an engine working on an Otto cycle and far below that of one working on a Carnot cycle. The differences between the practical and the Otto cycles are that processes a_1-a_2 and c_1-c differ as has just been mentioned and when both practical and Otto cycles are compared with the Carnot cycle it is noted that with the Carnot cycle energy is added and subtracted by heat at constant temperatures whereas in the other two cycles it is not. An Otto cycle is a useful standard of comparison for the practical cycle in that it represents more closely than the Carnot cycle what actually happens in a petrol engine.

The Otto efficiency is given as follows from equation (6.8) expressed in terms of energy per unit mass of working fluid,

$$\eta = \frac{w}{q_1}$$

$$= 1 + \frac{q_2}{q_1}$$

$$= 1 + \frac{c_v (T_c - T_{c_1})}{c'_v (T_{a_2} - T_{a_1})}$$

For a perfect gas c_v is constant so $c_v = c'_v$

then $$\eta = 1 + \frac{T_c - C_{c_1}}{T_{a_2} - T_{a_1}}$$

For a perfect gas undergoing the process $c - a_1$

$$\frac{P_c V_1}{T_c} = \frac{P_{a_1} V_2}{T_{a_1}} \text{ or } \frac{T_{a_1}}{T_c} = \frac{P_{a_1} V_2}{P_c V_1}$$

As this process is isentropic

$$P_c V_1^\gamma = P_{a1} V_2^\gamma \text{ or } \frac{T_{a1}}{T_c} = \left(\frac{V_1}{V_2}\right)^{\gamma - 1}$$

Let $r_V = V_1/V_2$ where r_V is called the **Volume ratio**

then $$\frac{T_{a_1}}{T_c} = (r_V)^{\gamma - 1}$$

Similarly for the process $a_2 - c_1$

$$\frac{T_{a_2}}{T_{cl}} = (r_V)^{\gamma - 1}$$

Then $$\frac{T_{a_1} - T_{a_2}}{T_c - T_{c_1}} = (r_V)^{\gamma - 1}$$

so $$\eta = 1 - \frac{1}{(r_V)^{\gamma - 1}} \qquad\qquad (23.1)$$

Equation (23.1) is a more realistic standard of comparison than Carnot's for the efficiency of a cycle using a reciprocating engine.

23.4 The Joule cycle

The cycle shown in Fig. 23.5 is approximately the same as the pressure-volume diagram of a gas engine and is known as the Joule or Brayton cycle in which the system that is for the short space of time enclosed in the engine undergoes the following four processes,

1-2   Isentropic compression from p_1 to p_2.
2-3   Addition of energy by heat at constant pressure causing an increase of temperature and volume. This is considered at constant pressure because it compares with the practical case in which combustion occurs relatively slowly as the fuel is put into the cylinder.
3-4   Isentropic expansion from p_2 to p_1.
4-1   Energy is withdrawn from the system at constant pressure while the temperature and volume fall.

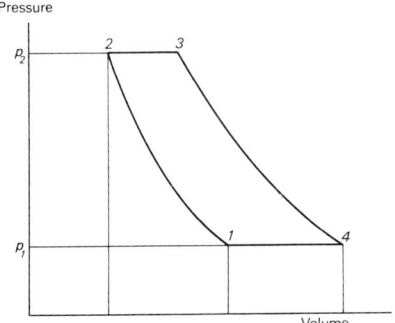

FIG 23. 5   The Joule cycle

Compared with the practical cycle this ideal Joule cycle differs in 2-3 and 4-1 in the same way that the practical petrol engine differs from the Otto cycle.

The efficiency of a practical gas or diesel engine is below the efficiency of the Joule cycle and far below one working on a Carnot cycle. However, the Joule cycle is a more useful standard of comparison than the Carnot cycle in that it represents more closely what actually happens in a gas engine.

The Joule efficiency is given as follows from equation (6. 8) expressed in terms of energy per unit mass of working fluid,

$$\eta = \frac{w}{q_1}$$

$$= \frac{q_1 + q_2}{q_1}$$

$$= \frac{c_p (T_3 - T_2) + c_p (T_1 - T_4)}{c_p (T_3 - T_2)}$$

$$= 1 - \left(\frac{T_4 - T_1}{T_3 - T_2}\right)$$

Using a similar substitution to that for the Otto cycle

$$= 1 - \left(\frac{1}{r_p}\right)^{\frac{\gamma - 1}{\gamma}} \tag{23. 2}$$

where r_p is the pressure ratio p_2/p_1. This is a more realistic standard of comparison than Carnot's for the efficiency of some practical cycles occurring in reciprocating engines.

23.5 The mixed cycle

The cycle shown in Fig. 23.6 is more nearly the same as the practical pressure-volume diagram of many reciprocating engines in which the system is for a short space of time enclosed in the engine and undergoes the following processes,

1-2   Isentropic compression from v_1 to v_2.
2-3   Addition of energy by heat, at constant volume causing a rise of temperature, and of pressure from p_2 to $p_{2''}$, during 2 to 2'. Followed by energy addition at constant pressure causing a further rise of temperature and a volume increase from $v_{2'}$ to v_3.
3-4   Isentropic expansion from v_3 to v_4.
4-1   Energy withdrawn from the system at constant volume while the temperature and the pressure fall from p_4 to p_1.

In this case the efficiency is, from equation (6.8), given by

$$\eta = 1 - \frac{r_p \, r_c{}^\gamma - 1}{r_v{}^{(\gamma-1)} (r_p - 1) + \gamma r_p (r_c - 1)} \tag{23.3}$$

where,

$$r_v = \frac{v_1}{v_2}, \qquad r_c = \frac{v_3}{v_2}, \qquad r_r = \frac{p_{2'}}{p_2}$$

This is a more realistic standard of comparison than the Otto or Joule cycle for the efficiency of many cycles using a reciprocating engine burning various fuels. A

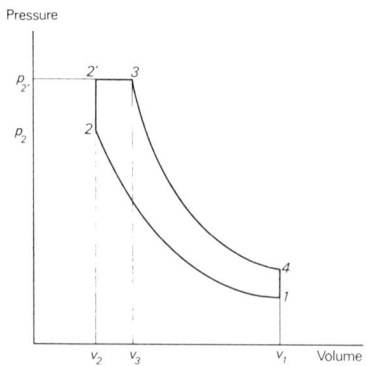

FIG 23.6   Mixed cycle

practical engine will give a pressure-volume diagram such as that shown in Fig. 23.7 where the idealised diagram has become rounded due to frictional and other losses.

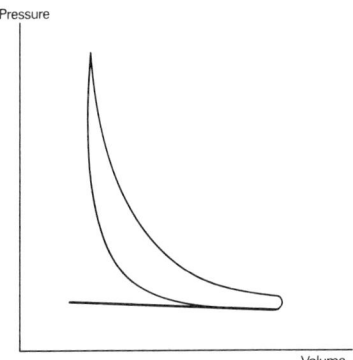

FIG 23.7   A practical pressure-volume diagram

23.6 The steam engine

The oldest type of reciprocating engine is a steam reciprocating engine. It should be noted that this is fundamentally different from the gas cycles in that the working fluid is a two-phase medium (liquid and vapour) whereas the other cases all consider a perfect gas. The steam engine does not perform the three functions of the gas turbine plant set out in section 22.1 as do petrol, gas and diesel reciprocating engines but performs the function only of the expander, between points 2 and 3 of the thermodynamic cycle of Fig. 6.3. If one considers the piston at inner dead centre ((a) of Fig. 23.1) the inlet valve has just opened and the steam admitted at the boiler pressure p_2, drives the piston out filling the swept volume with steam. The steam conditions in the cylinder will be the conditions between 2 and 2' in Fig. 23.8. The pressure of the steam between 2 and 2' is the same as the boiler or superheater pressure. During the change from 2 to 2' the piston has moved from its position at (a) in Fig. 23.1 to approximately its position at (b) in the same figure. At approximately this position of the piston the inlet valve IV of Fig. 23.2, closes and no more steam enters the cylinder. From this point the situation changes from one of steady flow to one of a closed system. It is for this reason that the position 2' is called **Cut-off**. Until the point 2' the mass of steam in the cylinder increases and after 2' it is regarded as a constant mass of steam. The working fluid expands in the cylinder as the piston moves from position (b) to position (c). The system in the cylinder continues to expand until the position of Fig. 23.1 (c) is reached, shown by point 3' in Fig. 23.8. When the piston is nearing outer dead centre, position (c) in Fig. 23.1, the exhaust valve opens, connecting the cylinder to the condenser. At this point of the valve opening the problem again changes from a closed system to a flow problem. The piston sweeps the steam out

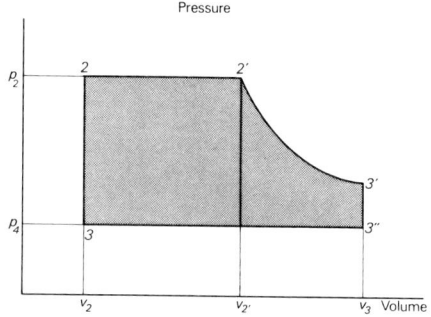

FIG 23.8   The steam reciprocating engine

of the cylinder into the condenser and the piston movement reduces the volume until the minimum volume, the clearance volume v_2, is reached.

The work done by the steam in expanding from volume v_2 to v_3 is given by

$$w = \int_2^3 p \, dv$$

$$= \int_2^{2'} p \, dv + \int_{2'}^{3'} p \, dv + \int_{3'}^3 p \, dv \tag{23.4}$$

If we apply the analysis of a system to this method we have to assume as with the Otto and Joule cycles that we are considering a fixed mass of gas going around the cycle. In this case we can write

$$w = p_2 \, (v_{2'} - v_2) + \int_{2'}^{3'} p \, dv + p_4 \, (v_3 - v_{3''}) \tag{23.5}$$

= area inside loop 2, 2′, 3′, 3″, 3, 2 shown crosshatched in Fig. 23.8.

23.7 Summary

When a petrol, gas or oil reciprocating engine—in fact any reciprocating engine that uses as a working fluid a fuel/air mixture irreversibly changed by combustion in the course of its circulation around the thermodynamic cycle—is used in a heat engine it may be said to take the place of the compressor, heater and expander of Fig. 6.8. Standards of efficiency for cycles using reciprocating engines such as the Otto, Joule and mixed cycles are more realistic as standards of performance in prcctice than that given by the Carnot cycle. When a steam reciprocating engine is used it replaces only the expander of a complete steam power plant. Efficiencies of the Otto, Joule and mixed cycles have been derived, and a graphical method for the work done by the steam in an idealised reciproacting engine has also been given.

23.8 Questions for the reader

Q.1. What is the fundamental difference between the Carnot, Joule, Otto and mixed cycles when compared with any practical reciprocating combustion cycle?

[In the practical case there is a change of mass and species]

Q.2. In the case of a practical combustion engine what are the main criteria to decide whether the p-v diagram (called the indicator diagram) will be similar to an Otto, Joule or mixed cycle?

[The method of fuel injection and the nature of the combustion process]

Q.3. Why is it difficult for a gaseous cycle to undergo a Carnot cycle?

[It is difficult to keep the temperature constant while adding and subtracting energy by heat—this implies changes of pressure and specific volume]

Q.4. Draw the T-s diagram for an Otto, Joule, mixed and Carnot cycles.

[See Fig. 23.9]

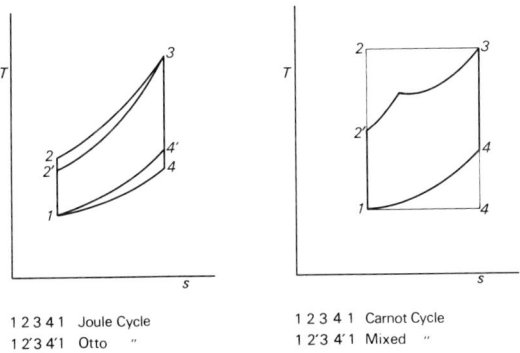

```
1 2 3 4 1   Joule Cycle          1 2 3 4 1   Carnot Cycle
1 2'3 4'1   Otto    "            1 2'3 4'1   Mixed    "
```

FIG 23.9   Answer to question 4

Q.5. Consider a series of engines working in an Otto cycle, sometimes called an air standard cycle, when air is used as the working fluid. For this case show how the efficiency varies with compression ratio. Take $\gamma = 1.4$.

[See Fig. 23.10]

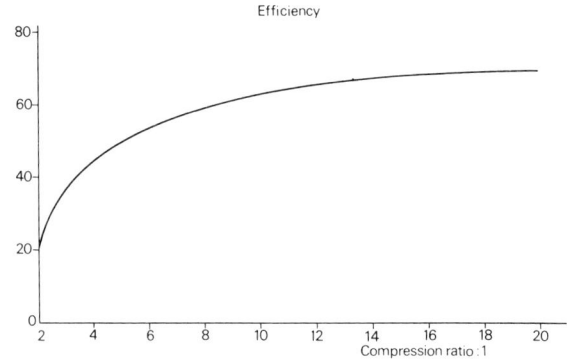

FIG 23. 10   Answer to question 5

Q. 6. A substance whose isentropic index $\gamma = 1 \cdot 2$ is expanded from volume v_1 to $5\ v_1$. Calculate the work done by the gas if the expansion is isothermal and reversible and if the substance is initially at 27°C.

[4016 kJ/kg mol]

Q. 7. Air is the working fluid in two engines one working on an Otto cycle and the other working on a Joule cycle. If the volume ratio of the first is 9, the pressure ratio of the second 9γ, and $\gamma = 1 \cdot 4$, calculate the efficiencies of these engines.

[Both 0·595]

24 A vapour power cycle

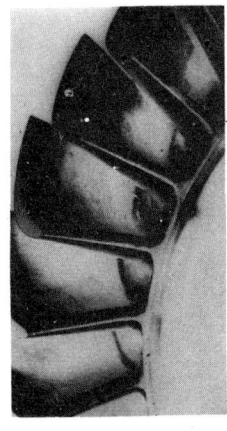

We now consider a change caused by changing the working fluid from a single-phase system, a gas, to a two-phase system, a liquid-vapour. The two types of plant, a gas power plant and a vapour power plant are compared.

24.1 A steam-turbine power cycle

A steam-turbine power plant is a form of heat engine (see Figs. 6.3 and 22.2). As has already been stated a heat engine is a system, comprising a working fluid undergoing a cyclic process, on the boundaries of which two transfers of heat energy q_1 and q_2 take place and one transfer of work energy w.

In Fig. 6.3 and again in Fig. 22.2 a power cycle is shown that consists of a working fluid—in the liquid and vapour states—circulating through the same series of components as did the gas in a gas power cycle. These are,

(a) A heat exchanger
(b) An expander
(c) Another heat exchanger
(d) A compressor.

These components are shown in Fig. 24.1 for a vapour power cycle when the components are better known as,

(a) A boiler or generator in which the fluid, frequently water, starts in the liquid state, takes in q_1 units of energy per unit mass of fluid, and finishes in the vapour state.
(b) A steam turbine in which the steam exchanges no heat energy with its surroundings (the process is therefore adiabatic) but does w_{xt} units of work energy per unit mass of the working fluid.

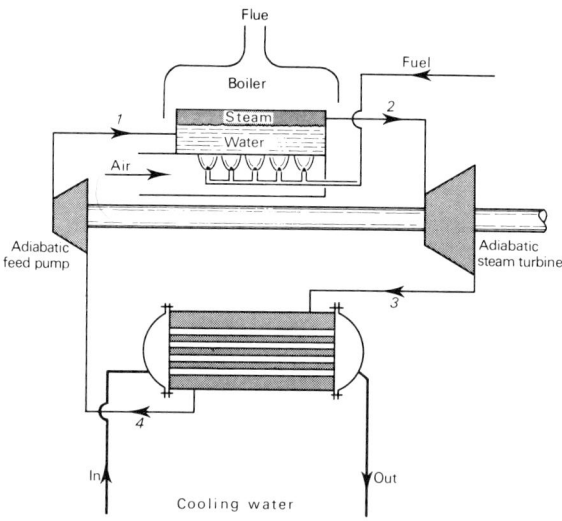

FIG 24. 1   A steam power plant

(c)   A condenser in which the fluid gives out q_2 units of heat energy per unit mass of working fluid. The working fluid would enter the condenser with a dryness of about 0·90 and would quite possibly leave the condenser as a saturated liquid.

(d)   A feed pump in which the fluid undergoes an adiabatic compression having w_{xp} units of work energy done on unit mass of working fluid.

24. 2   Components of a steam power plant

Figure 24. 1 shows the power plant of Fig. 6. 3 when water in the liquid and vapour states is used as the working fluid. The fluid moves round the closed cycle 1-2-3-4-1. Energy is supplied from the independent combustion of, say, a conventional fossil fuel or from a nuclear reactor. If it is conventional fuel the hot products of combustion pass around the boiler tubes to the flue. In passing around the colder surfaces of the tubes the hot products of combustion transfer energy by heat throught the walls of the tubes to the working fluid. The stored enthalpy of the working fluid is increased as it passes through the boiler tubes from point 1 where it is a liquid to point 2 where it is a vapour. From 2 the vapour passes through the adiabatic steam turbine, doing work on the turbine blades as it passes to point 3.

From 3 the working fluid enters the condenser passing across the outside of banks of cold condenser tubes, condensing as it goes, the water droplets falling to the bottom of the condenser after draining off the outside of the condenser tubes. The condenser tubes are usually kept cold by a supply of cooling water whose temperature is raised by an energy exchange through the walls of the tubes with the condensing vapour. The increase of the coolant's temperature as it passes through the condenser tubes will depend on the relative flow rates and storage capacities of the working fluid and of the cooling water. A feed pump then takes the working fluid, now a liquid, from the condenser, point 4, and compresses it to its original state at point 1 and feeds it again into the boiler. As the performance of the condenser can be seriously impaired by the presence of air, which may have leaked into the system or may have been released from solution in the water, an air extraction pump is used in conjunction with the feed pump or in the form of a separate pump.

24.3 Energy transfers in a vapour power cycle

If, as we did in Chapter 22 for the gas power cycle, we now examine more closely the energy transfers between the circulating working fluid and the surroundings we again have no difficulty in imagining where the work w goes to. It is used perhaps to drive electricity generators or ships—a future use might be their revival to drive automobiles. w is the sum of the two components w_{xt} and w_{xp}. w_{xt} is the work done by the vapour on the blades of the turbine and w_{xp} is that done on the condensed vapour by the feed pump. In fact (cf. equation (6.4))

$$w = w_{xt} + w_{xp} \tag{24.1}$$

The numerical value of w_{xt} is positive and of w_{xp} is negative, and as in the case of the gas turbine the work ratio is given by,

$$r_w = \frac{w_{xt} + w_{xp}}{w_{xt}} \tag{24.2}$$

This is a study of steadily flowing working fluid, in this case water, passing through a number of control volumes each of which encloses one of the components of the plant. Therefore, the steady-flow energy equation (8.6), applies to each control volume in terms of units of energy per unit time,

$$\dot{Q} - \dot{W}_x = \Delta\dot{H} + \Delta\dot{K} + \Delta\dot{Z} \tag{24.3}$$

or, per unit mass of working fluid per unit time,

$$\dot{q} - \dot{w}_x = \Delta\dot{h} + \Delta\dot{k} + \Delta\dot{z} \tag{24.4}$$

where $\Delta\dot{H} = \dot{M}\Delta h$, $\Delta\dot{K} = \dot{M}\Delta k$ and $\Delta\dot{Z} = \dot{M}\Delta z$.

These terms have already been discussed in Chapters 3 and 8—$\Delta\dot{H}$, $\Delta\dot{K}$ and $\Delta\dot{Z}$ are the changes per unit time of the enthalpy, kinetic and gravitational energies respectively, where $\dot{M}$ is the mass flow rate of the working fluid. Changes of $\dot{K}$ and $\dot{Z}$ are often sufficiently small to be neglected, and equation (24. 3) becomes

$$\dot{Q} - \dot{W}_X = \Delta \dot{H} \tag{24.5}$$

or, per unit mass of working fluid per unit time,

$$\dot{q} - \dot{w}_X = \Delta \dot{h} \tag{24.6}$$

Values of h, the enthalpy of the water, can be found from property tables. Because the state of the vapour in the cycle is often not far from the saturation line and from the critical point the fluid cannot be treated as a gas (see Chapter 18).

Equations (18. 2) and (18. 10) cannot be used for vapour as they are for a gas. Values of h, which have been calculated from experimental data and intermediate values interpolated with the aid of computers, are the subject of international agreement. These agreed values are tabulated in property tables or expressed graphically in the form of various charts, so that the values of h may be obtained in order to calculate efficiencies.

Components (Q and A)

Q. 1.   The boiler between points 1 and 2 of the cycle in Fig. 24. 1 works at a pressure of 200×10^5 N/m². Water comes from the feed pump at the rate of 2 kg/s and a temperature of 365·7°C. It enters the boiler with negligible velocity and energy is transferred to it by heat so that it leaves the boiler in a dry saturated state. Across the outer surfaces of the boiler shell 4 kg/s of combustion products pass, the temperature of the products on entering the boiler is 700°C. If the enthalpy lost by the combustion products equals that gained by the water, find the temperature of the combustion products leaving the boiler. (The exhaust products have $c_p = 1·101$ kJ/kg K.)

A. 1.   From the property tables we find,

Enthalpy of water entering the boiler = 1 827 kJ/kg
Enthalpy of water leaving the boiler = 2 411 kJ/kg

Rate of gain of enthalpy by the water in the boiler

$$\begin{aligned}
&= \dot{M}(h_2 - h_1) \\
&= 2 \times 584 \\
&= 1\,168 \text{ kJ/kg}
\end{aligned}$$

Rate of loss of enthalpy by products of combustion

$$\begin{aligned}
&= \dot{M}c_p(T_{out} - T_{in}) \\
&= 4·404\,(T_{out} - 700)
\end{aligned}$$

Therefore the temperature of the products of combustion leaving the boiler is given by

$$T_{out} = 700 - \frac{1\,168}{4\cdot404} = 435°C$$

It should be noted that these gases in the flue, at 435°C, are still at a higher temperature than the saturated steam at 200×10^5 N/m^2 which is at 365·7°C, so further energy could be extracted to superheat the steam.

Q. 2. In the same plant steam leaves the reversible and adiabatic turbine at 0·4 bar. What is the power output $\dot{W}_{xt}$ of the turbine? If the steam leaves the turbine at 200 m/s can we neglect the kinetic energy term?

A. 2. $-\dot{W}_{xt} = \dot{M}(h_3 - h_2)$ because $Q = 0$

From Question 1 we know $\dot{M} = 2$ kg/s and $h_2 = 2\,411$ kJ/kg. To find h_3 we must know the dryness, σ_3, at point 3. Because this process is adiabatic and reversible, we know $s_2 = s_3 = s_{31} + \sigma_3 s_{3lv}$ or from property tables

$$4\cdot928 = 1\cdot026 + \sigma_3\,6\cdot643$$

giving $\sigma_3 = 0\cdot587$

and so $h_3 = h_{31} + \sigma_3 h_{3lv}$
               $= 318 + 0\cdot587 \times 2\,318$
               $= 1\,679$ kJ/kg

Therefore $\dot{W}_{xt} = -2\,(1\,679 - 2\,411)$
               $= 1\,464$ kJ/s(kW)

The kinetic energy of 2 kg/s of steam travelling at 200 m/s

$$\dot{K} = \frac{2 \times 200 \times 200}{2 \times 1\,000} = 40 \text{ kJ/s (kW)}$$

which is small when compared with 1 464 kW total power output. (Note, the speed of this steam is high—it could be expressed as 720 km/h!)

24. 4 Temperature-entropy diagram

In the four components mentioned in this chapter and shown in Fig. 24. 1 the working fluid of the cycle undergoes four idealised energy transfer processes. These are,

(a) A reversible isothermal simple heat transfer process in the boiler, 1-2.

(b) A reversible adiabatic (isentropic) simple work transfer process in the turbine, 2-3.
(c) A reversible isothermal simple heat transfer process in the condenser, 3-4.
(d) A reversible adiabatic (isentropic) simple work transfer process in the feed pump, 4-1.

These processes, or changes of state, are shown in sequence on a temperature-entropy diagram in Fig. 24. 2. In practice a reversible isothermal heat transfer process is impossible to achieve, but because evaporation and condensation at constant pressure take place at constant temperature, an isothermal—although not reversible—process can be approached.

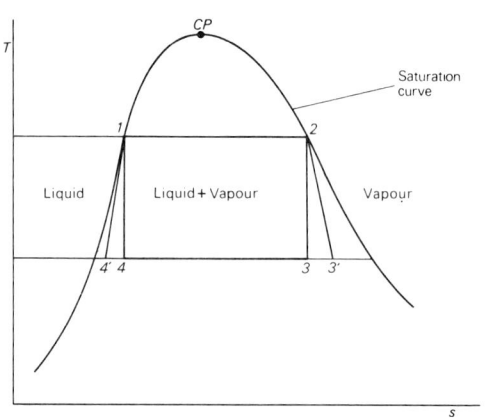

FIG 24. 2   The state path of a vapour in a vapour power plant

The properties of steam at the critical point are approximately 220×10^5 N/m^2 and 374°C. Steam in this region and also in the region of the satur tion line does not have the property of a gas described by the ideal gas rule (see equation (18. 1)) and so steam under these conditions is called a vapour. If, at the temperature the steam were to be used as a working fluid, the steam did behave as a gas then the cycle would have been a gas cycle and not a condensing vapour cycle. The gas power plant has been described in Chapter 22 and the cycle that the working fluid undergoes in a gas cycle is shown in Fig. 22. 4, whereas in this chapter the circulating fluid is working in such conditions that it condenses to a liquid within the cycle. Therefore for situations where steam is used as the working fluid and conditions are below 374°C and 220×10^5 N/m^2 the cycle is a vapour power cycle. Similarly the same type of cycle, a vapour power cycle, could be used if air was the working fluid if the conditions were below −141°C and 38×10^5 N/m^2 the critical point condition for air. Therefore, when air is used as the working fluid, because its behaviour approximates to that of a gas, the system used is that of a gas power plant. Unless a most inhospitable environment is found in outer space making possible temperatures well below −141°C it is unlikely that we shall ever have an air vapour power cycle!

It is interesting to study a T-s diagram on which a gas power and a vapour power cycle using the same working fluid and working in the same pressure range are shown together—this is done in Fig. 24. 3. The cycle 1_g-2_g-3_g-4_g is a gas cycle of the type discussed in Chapter 22, and 1_v-2_v-3_v-4_v-1_v represents a vapour cycle of the type discussed in this chapter. An important feature is the position on the diagram of each cycle relative to the saturation curve and the critical point.

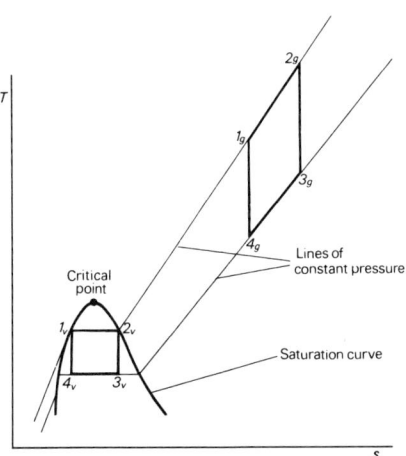

FIG 24. 3   Comparison of gas and vapour power cycles for a given working fluid

With regard to Fig. 24. 3 consider a fluid in common use, water. At about 220×10^5 N/m^2 the approximate values of c_p at 450, 550, 650 and 750°C are 4·2, 3·2, 2·8 and 2·7 kJ/kg K, from which it can be seen that it is only at temperatures higher than about 600°C that c_p is nearing a constant value. A constant value of c_p is one of the criteria of the fluid's being a perfect gas—the other being its obeying the ideal gas rule pv = RT (see equation (18. 1)). It follows that 600°C is the lowest temperature for a reservoir that can be used in what could be called a steam gas cycle working in such a way that its lowest pressure is 220×10^5 N/m^2.

24. 5 Isentropic efficiency

The change of state from 2 to 3 shown in Fig. 24. 2 is the state path of the working fluid in a vapour power plant as it passes through a reversible adiabatic, and therefore isentropic, turbine. This however is an ideal case because, although all turbines can be considered adiabatic, in practice they are not reversible, because of for example, the effect of friction. The actual path of the vapour as it passes through the turbine is not 2-3 but 2-3'. The vapour starts and finishes at the same pressures in the reverṣible process 2-3 as in the irreversible process 2-3'. The processes undergone are both adiabatic and in the ideal case 2-3 it is isentropic (no increase in the entropy, $s_2 = s_3$) but in the irreversible case the entropy increases (cf. the second law which was shown, so far as a gas is concerned to lead one to the conclusion that during an adiabatic process $\Delta s \geqslant 0$). So it is that the state path is from 2 to 3' where the final pressure is the same at 3 as it is at 3' but in going from 2 to 3' there is an increase of entropy.

Similarly during the compression 4-1 in the feed pump, the ideal reversible adiabatic process is shown by the line 4-1, but the irreversible adiabatic process is shown by 4'-1. During the process 4'-1 the entropy has increased while during the ideal process 4-1 the change is isentropic and the entropy $s_1 = s_4$.

In all the processes just mentioned $q = 0$ because all are adiabatic. Also, for the reversible adiabatic processes 2-3 and 4-1, $w_x = w_{xrt}$ in one case and $w_x = w_{xrp}$ in the other. If changes of kinetic and gravitational energy are negligible the steady-flow energy equation (8.13) becomes

$$w_{xrt} = -\Delta h = -(h_3 - h_2) \qquad (24.7)$$

and $\qquad w_{xrp} = -\Delta h = -(h_1 - h_4) \qquad (24.8)$

and, for the irreversible and adiabatic processes 2-3' and 4'-1,

$$w_{xt} = -(h_{3'} - h_2) \qquad (24.9)$$

$$w_{xp} = -(h_1 - h_{4'}) \qquad (24.10)$$

Equation (24.7) gives the work done in the course of an ideal isentropic expansion and (24.9) is the work done in the course of a practical irreversible expansion within the same pressure range. The ratio

$$\eta_I = \frac{w_{xt}}{w_{xrt}} = \frac{-(h_{3'} - h_2)}{-(h_3 - h_2)} \qquad (24.11)$$

is called the isentropic efficiency. Although it is not an efficiency within the definition given by equation (6.7) it is a ratio of the work energy given out by a practical irreversible turbine to the work energy given out by a reversible turbine when the fluid entering each turbine is in the same state and the pressure of the fluid leaving each turbine is the same.

Similarly for the adiabatic compressor the isentropic efficiency is given by

$$\eta_I = \frac{w_{xrp}}{w_{xp}} = \frac{-(h_1 - h_4)}{-(h_{1'} - h_4)} \qquad (24.12$$

As we mentioned in Chapter 22, it is conventional to have a different ratio for the turbine equation (24. 11) from that for the compressor, equation (24. 12).

24. 6 Summary

The heat engine using a vapour power cycle and its components has been described and compared with one using a gas power cycle. A T-s diagram is shown for an ideal vapour power plant and it is discussed how the practical case would differ from this ideal.

24. 7 Questions for the reader

Q. 1.   A steam power plant supplies wet vapour from a condenser at $0 \cdot 6 \times 10^5$ N/m^2 pressure to a feed pump at such a dryness that after reversible compression in the feed pump to 50×10^5 N/m^2 the fluid is saturated liquid. The water, as it passes through a boiler, has its dryness raised from that of saturated water, $\sigma = 0$, to dry saturated steam, $\sigma = 1$.

Sketch the T-s diagram for this cycle giving maximum and minimum values of T and s.

[See Fig. 24. 4]

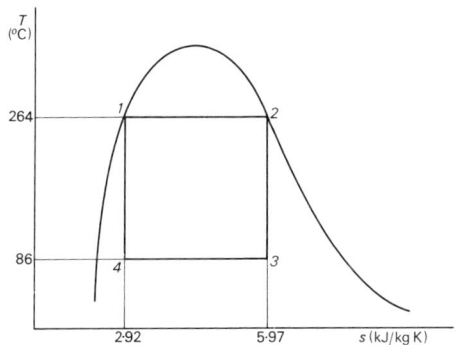

FIG 24. 4   Answer to question 1

Q. 2.   Noting that the cycle of Question 1 is operating on a Carnot cycle, what is the efficiency of the cycle?

[0·332]

Q. 3.   If the answer to Question 2 were calculated using the values of enthalpy obtained by calculation and from the property tables rather than from the temperatures, how would they differ?

[No difference, except from errors due to inaccuracy in the tabulated values, because the cycle is a perfect Carnot cycle]

Q. 4.   Consider the plant in Question 1 to be a practical plant when the turbine has an isentropic efficiency of 0·90 and the work ratio of the plant is 0·08. Calculate the efficiency of this plant.

[0· 308]

Q. 5.   Why is this efficiency lower than that for the plant described in Questions 1 and 2?

[Because in this practical case the turbine and compressor processes are irreversible]

Q. 6.   Draw a T–s diagram for the practical plant.

[See Fig. 24. 5]

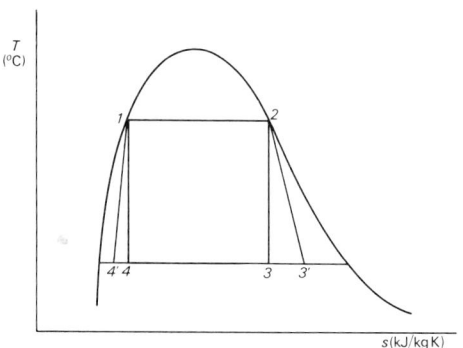

FIG 24. 5   Answer to question 6

Q. 7.   A steam turbine admits steam at 30×10^5 N/m^2 superheated to 420°C and discharges it at $0·20 \times 10^5$ N/m^2 with a dryness of 0·92. What is the turbine's isentropic efficiency?

$[w_{xrt} = 975; w_{xt} = 850$ kJ/kg$; \eta_I = 0·872]$

Q. 8.   In the plant of Question 7 the condenser discharges saturated water at $0·20 \times 10^5$ N/m^2. What is the plant's approximate efficiency?

$[q_1 = 3\,024$ kJ/kg$; \eta = 0·281]$

25

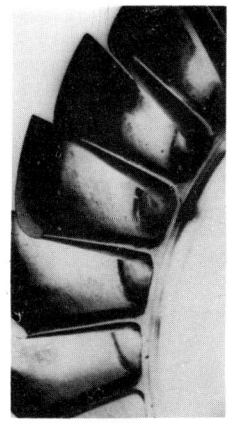

The Rankine cycle

The vapour power plant is now considered in a practical form where the working fluid undergoes a Rankine cycle. The practical Rankine cycle is then compared with the theoretical Carnot cycle.

25.1 The superheater

In Fig. 24.2 the thermodynamic cycle of a vapour power plant working on a Carnot cycle is shown by 1-2-3-4-1. One of the features of a Carnot cycle is that two of its processes, in this case processes 1-2 and 3-4, are isothermal as well as reversible. Although, in a practical boiler or condenser working at constant pressure, it is fairly easy to make processes 1-2 and 3-4 isothermal, reversibility is not achieved. This is because the products of combustion that heat the boiler tubes are at a temperature higher than that of the liquid, which is contrary to the rule for reversibility of a process involving heat transfer derived in Chapter 13 from the second law of thermodynamics. However, although not reversible, the two processes at least meet the isothermal requirement of Carnot.

The state of the vapour as it enters the turbine is represented by point 2 in Fig. 24.2, where it is dry saturated, but while it is in the turbine the state point moves from 2 to 3 and it can be seen in the figure that at 3 the steam is not dry saturated at all. In fact, as its state moves from 2 to 3, the vapour gets wetter until, at 3, there is a quantity of liquid in the turbine. This is a dangerous situation. At best the loose liquid, moving at high speed, can cause erosion of the turbine blades. At worst, given certain conditions, the drops of liquid can contribute to the breaking off of the blades. With some loss of efficiency the wetness of the vapour in the turbine can be reduced, or removed altogether, by superheating the vapour. The superheating process would be carried out at the boiler pressure but in a component called a superheater where the enthalpy of the vapour is increased by energy transfer by heat. So the vapour at point 2 in Fig. 25.1 would pass from the boiler process to the superheater where its temperature would be raised at constant pressure to point 2_s. If, from state 2_s, the vapour were expanded isentropically in the turbine its state would move along path 2_s-3 of Fig. 25.1 to state 3. In state 3 it would be drier than at the end of path 2-3 of Fig. 24.2 shown dotted in Fig. 25.1. The dotted path of Fig. 25.1 from 2 shows steam admitted in a dry saturated state to the turbine. Both the isentropic expansions from 2 and 2_s show ideal reversible turbine behaviour which is not obtained in practice. In practical terms the expansion 2-3' from the point 2 is shown in Fig. 24.2, and the equivalent practical case for the expansion from 2_s is the path 2_s-3'. The stea at state points 3 and 3' of Fig. 25.1 is drier than it was at state point 5 of Fig. 25.1 and so the situation with regard to the presence of liquid is less dangerous.

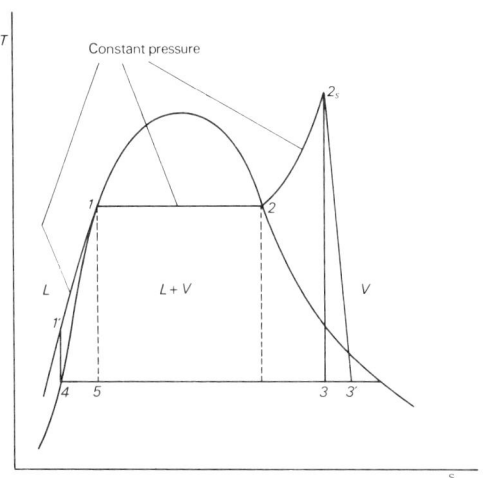

FIG 25.1   The Rankine cycle T-s diagram

25.2 The flow diagram

The flow diagram for an elementary vapour plant is shown in Fig. 24. for the working fluid going around the thermodynamic cycle 1-2-3-4-1 or 1-2-3'-4'-1 of Fig. 24.2. The flow diagram with a superheater as one of the plant's components is shown in Fig. 25.2. The vapour in the turbine after leaving the superheater, as its state moves along the path 2_s-3 in an ideal turbine or 2_s-3' for a practical turbine, is for some time superheated and for the rest of the time drier than it was without the superheater (the dotted path of Fig. 25.1). The possible gain, by making the process safer and reducing the chances of blade damage, has been made at the cost of a lower efficiency. This is because q_1 is transferred during a process that is no longer an isothermal process with the fluid at the maximum available temperature. The process 1-2 was isothermal at the maximu

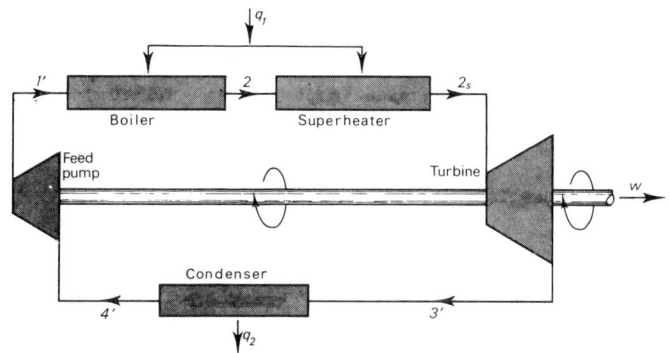

FIG 25. 2   The Rankine cycle; Flow diagram

available temperature but process 1-2$_s$ is not, in that the temperature at T$_s$ is higher than that at either 1 or 2.

It was shown in an example in Chapter 24, with reference to the boiler, that the temperature of the products of combustion was higher than the temperature of the working fluid passing through the boiler. This fact can be used to raise the temperature of the working fluid—to superheat the vapour. The superheater may be an integral part of the boiler or it may be a separate unit with an independent energy supply.

25. 3 The feed pump

In Fig. 24. 2 the feed pump is shown to have finished its isentropic compression at point 1–that is to say at a point where the working fluid entering the boiler is saturated liquid. The feed pump takes in fluid at point 4 and discharges it into the boiler at point 1. At state 4 of Fig. 24. 2 the working fluid still has a significant dryness and the volume of the fluid per unit mass may be so large that the size of a feed pump of sufficient capacity to handle the working fluid would be considerable. This would require a large amount of work energy to be supplied to the compressor. For instance, the dryness at point 4 might be 0·290, and the specific volume of steam at a condenser pressure of $0·10 \times 10^5$ N/m^2 would be 4·25 m^3/kg whereas if condensation had been continued at $0·10 \times 10^5$ N/m^2 pressure until the fluid were wholly liquid the specific volume would be 0·00101 m^3/kg.

The line 1-2-2$_s$ of Fig. 25. 1 is a line of constant pressure which, if continued backwards from 1, falls sharply down, following the curve 1-1'. It has been found more economic to let the working fluid fully condense in the condenser before letting it pass on to the feed pump. So condensation goes on after point 4 of Fig. 24. 2 and continues to point 4 of Fig. 25. 1 where the fluid is saturated liquid at the condenser pressure and the fluid's specific volume is very small. If in this state the liquid is admitted to the feed pump little work need be done because liquid water is so very small. After the working fluid is fully condensed to point 4 on Fig. 25.1 it is compressed in the feed pump to point $1'$ where it is at boiler pressure. It is in fact in state $1'$ when it is fed into the boiler—that is at a lower temperature than if it were a saturated liquid at 1. In the boiler it is heated from state $1'$ to state 1 and so on to state 2 and further to 2_s. The fact that the energy i. added over a range of temperature $1'$-1, as it is for 2-2_s, represents a further decrease in efficiency, because it further deviates from the reversible isothermal process. In Carnot's perfect cycle the energy would be added isothermally at the maximum temperature, 2_s, but in this more realistic cycle it is added over the range $1'$-1-2-2_s.

Nevertheless the saving in the size of the feed pump, the use of the superheater, and the prevention of blade damage, makes the cycle a more practical proposition. The otherwise waste energy of the combustion process is used to heat the steam from a lower temperature and less work energy is required to compress the fluid in the feed pump. A result is that more work is obtained from the turbine but that overall the efficiency is lower. In addition the increased dryness of the steam in the turbine due to superheating makes it possible to use practical materials for the manufacture of turbine blades with less risk of excessive blade damage.

25.4 The Rankine cycle

The cycle $1'$-1-2-2_s-3-4-$1'$ of Fig. 25.1 is called the Rankine cycle, and the cycle is preferred to Carnot's as a criterion for vapour power plants because it is more realistic. The quantity of heat energy q_1 comprises two parts, one part transferred to the working fluid in the boiler and the other in the superheater. Therefore, from the steady-flow energy equation,

$$q_1 = (h_{2s} - h_{1'}) \qquad (25.)$$

Also, from the steady-flow energy equation applied to the turbine and to the feed pump,

$$w_{xrt} = -(h_3 - h_2) \qquad (25.)$$

$$w_{xrp} = -(h_{1'} - h_4) \qquad (25.)$$

and, for the condenser,

$$q_2 = (h_4 - h_3) \qquad (25.)$$

(Note: in each case it has been assumed that Δk and Δz terms are negligible—in any practical case it would be necessary to demonstrate that this was so.)

The Rankine efficiency, the criterion of efficiency used for vapour power plants, is,

$$\eta_{Rank} = \frac{w_r}{q_{r1}} \qquad \text{from equation (6.8)}$$

$$= \frac{w_{xrt} + w_{xrp}}{q_{r1}}$$

$$= \frac{-(h_3 - h_{2s}) - (h_{1'} - h_4)}{(h_{2s} - h_{1'})} \tag{25.5}$$

The Carnot efficiency in the same case would be,

$$\eta_r = \frac{T_{2s} - T_4}{T_{2s}} \tag{25.6}$$

It is often the case that the work done in a feed pump, given by equation (25.3) as,

$$w_{xrp} = -(h_{1'} - h_4)$$

is so small that it can be taken as zero, in which case

$$h_{1'} \simeq h_4 \tag{25.7}$$

If a more accurate value of w_{xrp} than the assumed zero is required this is not easily obtained because $h_{1'}$ does not lie on the saturation line and so cannot be read from the property tables. Therefore, instead of finding the feed pump work directly from values of $h_{1'}$ and h_4, it is often preferable to use the fact that for an incompressible fluid undergoing an isentropic process

$$
\begin{aligned}
w_{xrp} \quad &= -(h_{1'} - h_4) \\
&= \text{force used by piston of feed pump to drive the water into} \\
&\quad \text{the boiler} \\
&= (P_{1'} - p_4)v_{14} \quad \text{because } v_{14} \simeq v_{11'} \\
(h_{1'} - h_4) &= v_{14}(p_{1'} - p_4)
\end{aligned}
\tag{25.8}
$$

What we are calculating in equation (25.8) is the work done by the piston of the feed pump while pushing unit mass of liquid at the temperature of point 4 (hence the use of v_{14}) into a boiler against the boiler pressure of point 1' (hence the use of $p_{1'}$).

The Rankine cycle (Q and A)

Q.1. The higher pressure of a steam plant is 40×10^5 N/m^2 and the lower pressure 0.30×10^5 N/m^2. Calculate the work that would be done in the feed pump if

(a) The fluid were discharged from the condenser into the pump in such a condition at the lower pressure that after isentropic compression its condition was that of a saturated liquid, and

(b) The condenser is used to make the fluid a saturated liquid at the lower pressure so that isentropic compression occurs with the fluid totally in the liquid state.

A. 1.   (a) For the path 5-1 in the diagram, Fig. 25. 1.

$$s_5 = s_1 = 2 \cdot 797 \text{ kJ/kg K}$$
$$= s_4 + \sigma_5 \, (s_6 - s_5) \text{ [equation 15. 3]}$$
$$= 0 \cdot 944 + \sigma_5 \, 6 \cdot 823$$

giving $$\sigma_5 = 0 \cdot 272$$
$$h_5 = h_4 + \sigma_5 \, (h_6 - h_4)$$
$$= 289 + 0 \cdot 272 \times 2336$$
$$= 924 \text{ kJ/kg}$$
$$w_{xrp} = -(h_1 - h_5) = -(1\,087 - 924) = -163 \text{ kJ/kg}$$

(b) For the path $4 - 1'$, here we apply equation (25. 8). The boiler pressure $p_{1'} = 40 \times 10^5 \text{ N/m}^2$ and the condenser pressure $p_4 = 0 \cdot 30 \times 10^5 \text{ N/m}^2$. At this pressure the specific volume of the water is $0 \cdot 00102 \text{ m}^3 /\text{kg}$

so $$w_{xrp} \simeq \frac{-0 \cdot 00102 \times 40 \times 10^5 \text{ kJ/kg}}{10^3}$$

$$\simeq -4 \cdot 08 \text{ kJ/kg}$$

By   By comparing $-4 \cdot 08 \text{ kJ/kg}$ with the answer for part (a) of -163 kJ/kg it can be seen that less work energy from the turbine is required to drive the feed pump when the working fluid is allowed to fully condense at the condenser pressure.

Q. 2. Compare the Rankine with the Carnot efficiency for the plant of Question 1 if, before entering the turbine, the steam were superheated to 350°C.
A. 2.   From the steam property tables, $h_{2s} = 3\,094 \text{ kJ/kg}$, and

$$s_{2s} = 6 \cdot 584 \text{ kJ/kg K}$$
$$= s_{l3} + \sigma_3 \, s_{lv3}$$
$$= 0 \cdot 944 + \sigma_3 \, 6 \cdot 823$$

giving $$\sigma_3 = 0 \cdot 826$$
so $$h_3 = 289 + 0 \cdot 826 \times 2\,336$$
$$= 2\,219 \text{ kJ/kg}$$

Suppose $$w_{xrp} = 0 \text{ (see A. 1. (b))}$$
then $$h_{1'} = h_4 = 289 \text{ kJ/kg}$$

From equation (25. 5)

$$\eta_{\text{Rank}} = -\frac{(2\,219 - 3\,094) - (0)}{3\,094 - 289}$$

$$= 0 \cdot 312$$

From equation (25. 6)

$$\eta_{\text{r}} = \frac{[(350 + 273) - (69 + 273)]}{[350 + 273]}$$

$$= 0 \cdot 452$$

25. 5 The enthalpy–entropy diagram

The property tables, and diagrams based on them, such as the temperature-entropy diagrams of Figs. 17. 3 and 25. 1, can be used when obtaining values of properties of a working fluid particularly a liquid or vapour. In particular the enthalpy-entropy diagram, such as that mentioned in Ref. (4), of which a part is shown in Fig. 25. 3 is useful.

When steam leaves the boiler in, say, state 2 shown by point 2 in Fig. 25. 1 it enters a superheater in which its pressure remains the same but its temperature—which was at the saturation temperature—is raised to the temperature of point 2_s in the figure. Its enthalpy h_{2s} and its entropy S_{2s} can be read from the chart. In the Rankine cycle the steam at 2_s enters the isentropic turbine where its pressure falls to the condenser pressure and, where the line of constant entropy through point 2_s crosses the line of the pressure in the condenser, there is point 3. The enthalpy h_3 can be read from the chart and the work that would be done isentropically in the turbine can be found from equation (25. 2).

$$w_{\text{xrt}} = -(h_3 - h_2)$$

If the isentropic efficiency of the turbine has a known value η_I the work actually done in the turbine can be found from equations (25. 2) and (22. 9) and is

$$w_{\text{xt}} = \eta_I w_{\text{xrt}}$$
$$= -(h_{3'} - h_{2s}) \tag{25.9}$$

As all other values are known in equation (25. 9) except $h_{3'}$ the equation can now be used to find $h_{3'}$ and the state point 3' which lies where the line of condenser pressure crosses the horizontal enthalpy line showing the enthalpy at point 3'.

25. 6 Example of use of h–s diagram

If the pressure in the boiler is 6×10^5 N/m^2 the steam leaves the boiler in the state shown by point 2 when it enters the superheater. In the superheater the pressure remains at 6×10^5 N/m^2 but the temperature rises to 220°C.

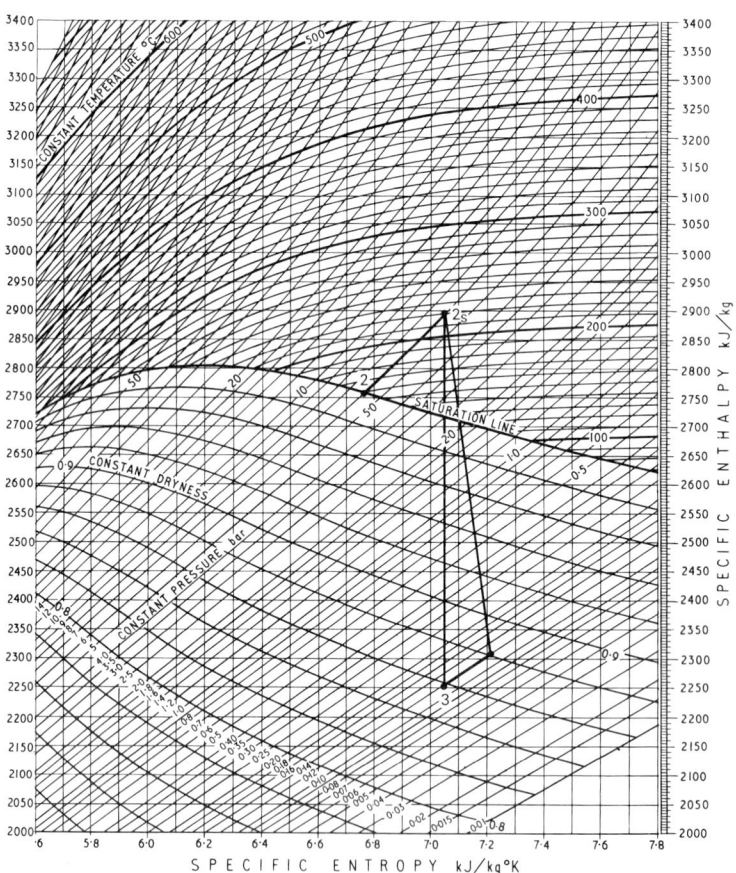

This is the state point 2_s in Fig. 25.3 and is situated at the point at which the pressure line of 6 bars crosses the temperature line of 220°C. The steam enters the turbine in state 2_s and, if the process in the turbine were reversible, would leave it in state 3. However, if the turbine has an isentropic efficiency of 0·92 it leaves the turbine in state 3'. In the following calculation the practical Rankine cycles and the ideal Carnot efficiencies are calculated for this particular case. We can begine by taking, from Fig. 25.3, values of enthalpy, entropy, and, where relevant, dryness at state points $2, 2_s$ and 3.

$$h_2 = 2\,760 \text{ kJ/kg} \qquad s_2 = 6·77 \text{ kJ/kg K} \qquad \sigma_2 = 1·00$$
$$h_{2s} = 2\,895 \text{ kJ/kg} \qquad s_{2s} = 7·07 \text{ kJ/kg K}$$
$$h_3 = 2\,255 \text{ kJ/kg} \qquad s_3 = 7·07 \text{ kJ/kg K} \qquad \sigma_3 = 0·858$$

For an isentropic expansion in the turbine, the work done would be

$$w_{xrt} = -(h_3 - h_{2s})$$
$$= -(2\,255 - 2\,895)$$
$$= 640 \text{ kJ/kg}$$

If, however, the turbine had an isentropic efficiency of 0·92, then from equation (24.9)

$$w_{xt} = 0·92 \times 640$$
$$= 589 \text{ kJ/kg}$$
$$= -(h_{3'} - h_{2s}) \qquad (25.9)$$

substituting in the value of h_{2s} we get,

$$h_{3'} = 2\,895 - 589$$
$$= 2\,306 \text{ kJ/kg}$$

At a pressure of $0·12 \times 10^5 \text{ N/m}^2$ the point at which the vapour's enthalpy is 2 306 kJ/kg can be found on Fig. 25.3 with an accuracy that is usually sufficient for engineers. The point 3' is marked on the figure and we see that at this point $\sigma_{3'} = 0·880$, compared with the point 3 where $\sigma_3 = 0·858$.

If a value for h_4, of water at $0·12 \times 10^5 \text{ N/m}^2$ in the saturated liquid state, is required, or the values of any properties of a vapour when the dryness is low, the property tables must be used. If p_4 is $0·12 \times 10^5 \text{ N/m}^2$ it can be seen from the tables that h_4 is 207 kJ/kg. Therefore

$$q_1 = (h_{2s} - h_{1'}) \text{ from equation (25.1)}$$
$$\simeq (h_{2s} - h_4) \text{ from equation (25.7)}$$
$$= (2\,895 - 207)$$
$$= 2\,688 \text{ kJ/kg}$$
$$w = (w_{xt} + w_{xp}) \text{ from equation (24.1)}$$
$$\simeq w_{xt}$$
$$= -(h_{3'} - h_{2s}) \text{ from equation (25.9)}$$
$$= -(2\,306 - 2\,895)$$
$$= 589 \text{ kJ/kg}$$

q_2 is given either by $(q_1 + q_2) - w = 0$

$$q_2 = -2\,099 \text{ kJ/kg}$$

or

$$q_2 = (h_4 - h_{3'}) \text{ from equation (25.4), substituting } h_{3'} \text{ for } h_3$$
$$= (207 - 2\,311)$$
$$= -2\,099 \text{ kJ/kg}$$

For the case in which the turbine is isentropic,

$$\eta_{\text{Rank}} = \frac{w_{rt}}{q_1} = \frac{640}{2\,688} = 0 \cdot 238$$

but for the case in which the turbine has an isentropic efficiency of $0 \cdot 92$

$$\eta = \frac{w_t}{q_1} = \frac{589}{2\,688} = 0 \cdot 219$$

and the Carnot efficiency for the same plant would be given by,

$$\eta_r = \frac{T_{\text{Max}} - T_{\text{Min}}}{T_{\text{Max}}}$$
$$= \frac{493 - 332}{493}$$
$$= 0 \cdot 327$$

However the Rankine efficiency η_{Rank} is the efficiency of an engine of which the processes more nearly correspond with the processes of a real vapour engine. It is therefore a better standard of comparison than the Carnot efficiency η_r.

Steam plant (Q and A)

Q. 1. In a steam power plant working on a Rankine cycle between. 20×10^5 and $0 \cdot 02 \times 10^5$ N/m^2 what is the specific enthalpy of the steam as it enters the super-heater?

A. 1. The steam enters the superheater in state 2 shown on the T–s diagram.

From the enthalpy-entropy chart,

$$h_2 = 2\,800 \text{ kJ/kg}$$

Q. 2. The temperature of the steam in the same plant is raised at constant pressure in the superheater to 400°C. What is q_1 for the whole plant?

A. 2.   From the h-s diagram

$$h_{2s} = 3\,245 \text{ kJ/kg}$$

and from the property tables

$$h_4 = 73 \text{ kJ/kg}$$
$$= h_{1'} \text{ from equation (25.7)}$$

so

$$q_1 = (h_{2s} - h_{1'}) \text{ from equation (25.1)}$$
$$= (3\,245 - 73)$$
$$= 3\,172 \text{ kJ/kg}$$

Q. 3.   In the same plant the isentropic efficiency of the turbine is 0·88.

What is the useful work energy output of the whole plant?

A. 3.

$$w_{xrt} = -(h_3 - h_{2s}) \text{ from equation (25.2), substituting } h_{2s} \text{ for } h_2$$
$$= -(2\,060 - 3\,245) \text{ from Fig. 25.3}$$
$$= 1\,185 \text{ kJ/kg}$$
$$w_{xt} = 1\,185 \times 0·88 = 1\,043 \text{ kJ/kg}$$
$$w_{xrp} = -(h_{1'} - h_4) \text{ from equation (25.3)}$$
$$= 0 \text{ from the assumption made in section 25.5, Answer 1.}$$

The useful work output of the whole plant is,

$$w = w_{xt} + w_{xp}$$
$$= 1\,043 \text{ kJ/kg}$$

Q. 4.   What heat energy is removed, from the working fluid, in the condenser?

A. 4.   Using $(q_1 + q_2) - w = 0$

$$q_2 = -3\,172 + 1\,043 = -2\,129 \text{ kJ/kg}$$

Q. 5.   What are the Rankine and the actual efficiencies?

A. 5.

$$\eta_{Rank} = \frac{w_{rt}}{q_1} = \frac{1\,185}{3\,172} = 0·373$$

from equation (25.5) and

$$\eta = \frac{w_t}{q_1} = \frac{1\,043}{3\,172} = 0.332$$

25.7 Summary

Methods of improving the usefulness of a vapour power plant are studied firstly by the introduction of a superheater and secondly by fully condensing the working fluid to saturated liquid at condenser pressure. These two features are imposed on the simple cycle discussed in Chapter 24.

25.8 Questions for the reader

Q.1. What is the work ratio for the two plants (see Fig. 25.4)

    (a) operating on cycle 1-2-3-5-1, and
    (b) operating on cycle 1'-1-2-3-4-1'.

    [0·78, 0·99]

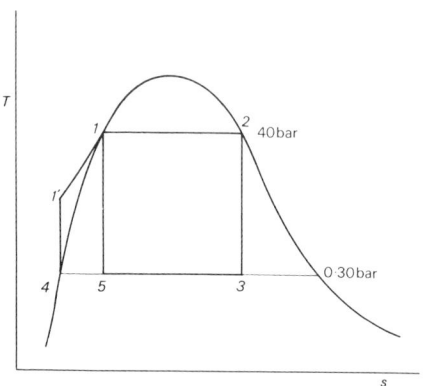

FIG 25.4   Cycle of question 1

Q.2. If a steam plant were operating on a Rankine cycle between a boiler pressure of 200 bar and a condenser pressure of 0·02 bar, how much work energy per unit mass of working fluid may be done on the feed pump?

    [21 kJ/kg]

Q.3. If the feed pump of Question 2 only had an isentropic efficiency of 0·5 how much work would be done on the working fluid?

    [42 kJ/kg]

Q.4. A vapour power plant operating on a Rankine cycle uses H_2O as its working fluid. The boiler is such that the temperature of the wet vapour is 179·9°C. After leaving the boiler the steam is superheated to 400°C. The turbine which is isentropic discharges into a condenser at 60·1°C. Find the efficiency of the plant.

[0·266]

Q. 5. In a simple steam-turbine plant operating on a Rankine cycle, the steam enters a reversible and adiabatic turbine at 25 bar and 450°C, and the condenser pressure is 0·02 bar. Calculate the efficiency of the plant.

[0·388]

Q. 6. If in Question 5 the water leaving the feed pump is a saturated liquid at boiler pressure, what is the efficiency of this plant?

[0·532]

Q. 7. Saturated water, assumed incompressible, leaves a condenser at 75°C and is delivered isentropically by a feed pump to a boiler at 30 bar. After leaving the boiler the steam is superheated to 400°C. The turbine has an isentropic efficiency of 0·80. Find the work done in the turbine, and the dryness of the working fluid leaving the turbine.

[700 kJ/kg, 0·95]

Q. 8. For the following Rankine cycle, calculate the efficiency and the work ratio, assuming the dryness at 3 is 0·90.

[0·346—see Fig. 25. 5]

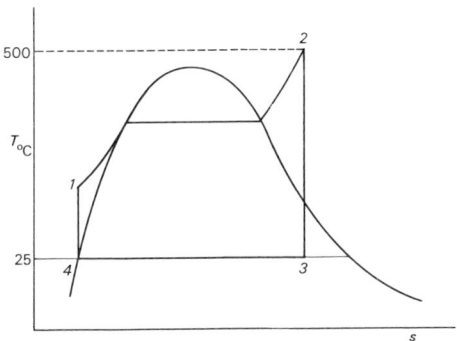

FIG 25. 5   Cycle of question 8

26 Heat pumps and refrigerators

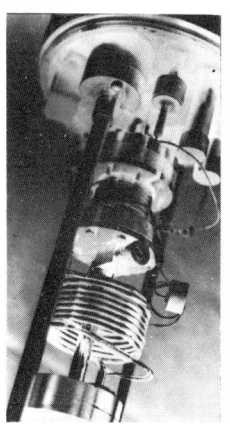

A reversed heat engine is considered in more detail and as these plants usually work on a reversed vapour power cycle the properties of a suitable working fluid are listed.

26.1 A reversed heat engine

A reversed heat engine is shown in Fig. 6.4 and again in Fig. 26.1; the only difference between these is that the first shows quantities Q_1', Q_2' and W', and the second shows the same energies q_1', q_2' and w', expressed in terms of units of energy per unit mass of the working fluid undergoing a thermodynamic cycle inside the reversed engine. It is in Fig. 26.1 that the engine, instead of doing work

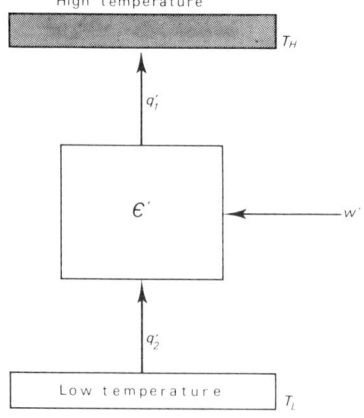

FIG 26.1   A reversed heat engine

energy w on the surroundings as it would if it were forward working, is having work energy w' done on it by the surroundings. The energy being transferred into the engine is w' by work and q_2' by heat while the energy q_1' is being transferred out of the engine by heat. These are energy exchanges between the working fluid of the reversed engine and the engine's surroundings. The circulation of the working fluid inside the engine is most easily seen in Fig. 26.2. Figure 26.2 is a flow diagram of a practical vapour heat pump.

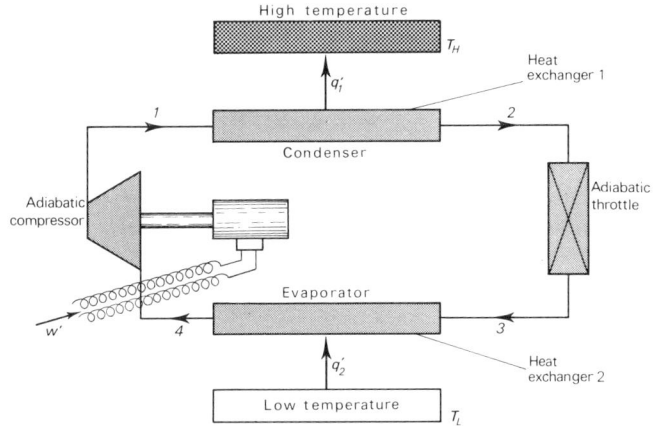

FIG 26.2   Flow diagram of a reversed heat engine

26.2 The components in the cycle

The vapour at point 1 of the cycle shown in Fig. 26.2, 26.3 and 26.4 would be at its highest temperature T_1, higher—as shown in Fig. 26.3—than the temperature T_H of the high-temperature reservoir. It would most probably be dry saturated vapour at point 1 as seen in Figs 26.3 and 26.4 which show typical T-s and h-s diagrams respectively for a reversed vapour engine. As the vapour passes through the condenser in Fig. 26.2 it gives up energy to the high-temperature reservoir and so the vapour, if the pressure in the condenser is constant, gets wetter until, when the fluid leaves the condenser at point 2, it is perhaps a saturated liquid as shown in Fig. 26.3. Because the fluid has throughout the process been a wet vapour, and because the process has taken place at constant pressure, the temperature at point 2 is the same as that at point 1. The energy given up during the process 1-2 is the enthalpy of condensation h_{lv} relevant to that saturation pressure. If we assume dry saturated vapour enters the condenser at point 1 of Figs. 26.2, 26.3 and 26.4 and leaves the condenser as a saturated liquid at point 2, applying the steady-flow energy equation to the process undergone by the fluid, we get

$$q_1' = (h_2 - h_1) \tag{26.1}$$

neglecting kinetic and gravitational energies.

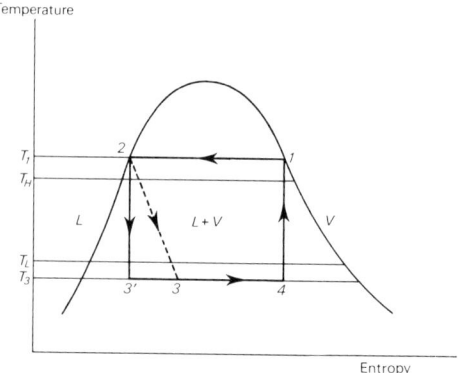

FIG 26.3   Thermodynamic cycle of a reversed heat engine

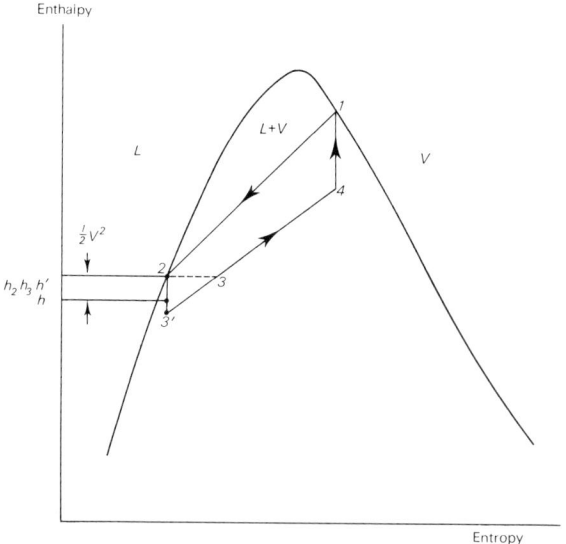

FIG 26.4   Thermodynamic cycle of a reversed heat engine

In state 2 the working fluid is probably a saturated liquid in which condition it enters, in a reversible cycle, an adiabatic nozzle such as that shown in Fig. 26.5. So a reversible expansion process is taking place within the nozzle. For a reversible expansion to occur the rate of convergence and divergence in the nozzle must be such that the flow of the fluid is at all times laminar, and for the same reason the walls of the nozzle have to be not only non-conducting but also

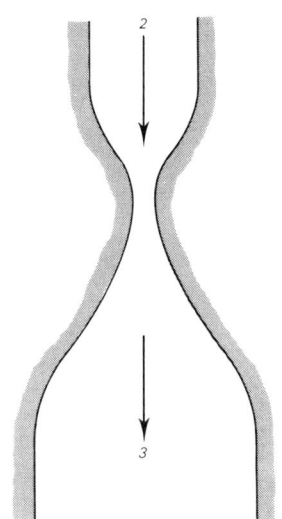

FIG 26.5   A convergent-divergent nozzle

smooth. In such a case $q = 0$, $w_{xr} = 0$, and in normal circumstances $\Delta z = 0$. The steady-flow energy equation would become,

$$0 = \Delta h + \Delta k$$

or $$0 = (h - h_2) + (k - k_2) \tag{26.2}$$

where h and k are specific enthalpy and kinetic energy at any cross-section of the nozzle. That is to say

$$h + \frac{v^2}{2} = \text{constant} = h'$$

or in other words at all cross-sections the stagnation enthalpy h' (see section 9.2 for definition) has the same value. The process that takes place is an adiabatic process and it is stated in section 15.2 that a reversible adiabatic process is isentropic. This demands that, for the reversible expansion in the adiabatic nozzle,

$$\Delta s = 0$$

or $$0 = s_3' - s_2 \tag{26.3}$$

It should be understood that equations (26.2) and (26.3) apply to a fluid going through a nozzle where the process is both adiabatic and reversible. The state path of such a process, starting with the fluid in a saturated liquid state at 2, in terms of enthalpy, entropy and of kinetic energy is shown by the line 2-3' in Fig. 26.4.

In the last paragraph we have been discussing a reversible adiabatic nozzle. In the practical cycle for a heat pump and vapour refrigerator the nozzle is not reversible and equation (26.3) does not apply. Also the kinetic energy changes are small, when vapour is used, compared with the changes in enthalpy and so equation (26.2) becomes

$$\Delta h = 0$$

or $$h_3 = h_2$$

This signifies that the process is one in which the enthalpy of the working fluid before and after passing through the nozzle is the same, although the pressure and therefore the temperature will have changed. It is worth noting that we only know about the end states of the process, $h_3 = h_2$. It has nothing to say—as has equation (26.2)—about the enthalpies at any cross-section of the nozzle. There will be a change in the dryness of the working fluid. Figure 26.3 shows how the temperature and dryness would vary. The state path 2-3 in Figs. 26.3 and 26.4 are shown by lines of dashes to indicate that the processes are only constant enthalpy as far as the end states of the process are concerned. During the process there will be high velocities hence high kinetic energies at different cross-sections and at times the value of the kinetic energy would not be negligible. Also in practice there will be exchanges of energy between the wall of the nozzle and the fluid but when the whole process is considered the values of q and w_x will be negligible.

The adiabatic process taking place in a reversible nozzle is a constant entropy process and the reduced form of the steady-flow energy equation (26.2) applies, shown by state paths 2-3' in Figs. 26.3 and 26.4. If the process is a practical process then the nozzle is not reversible, and is called a throttle in this case the only end states of the process are related by knowing they have the same enthalpy, shown by dashed lines joining the end states 2-3 in Figs. 26.3 and 26.4.

After passing through the throttle the working fluid enters a zone of low pressure at point 3 of Figs. 26.2 and 26.3. In changing from state 2 at high pressure to state 3 at low pressure it can be seen from Fig. 26.3 that the fluid has become drier at the lower pressure. The lower pressure is so selected by the designer of the plant that T_3 is lower than T_L, the temperature of the low-temperature reservoir so that energy may be transferred from the low-temperature reservoir into the working fluid as shown in Fig. 26.2. This additional energy makes the fluid drier while its pressure remains constant and, because the pressure in the evaporator is kept constant, the temperature of the wet fluid vapour remains the same and the state path is the isothermal line 3-4 in Fig. 26.3. The process of making a fluid drier at constant pressure is called evaporation and for this reason this component is called the evaporator. During the process 3-4 no work is done so the steady-flow energy equation for the process gives

$$q_2 = (h_4 - h_3) \tag{26.4}$$

The final component in the cycle is the compressor which takes the working fluid from state 4 back to state 1. State 4 is a fairly dry state at low pressure, and the process 4-1 involves adiabatic compression which further dries the working fluid to state 1 which is at a higher pressure. If the compressor works reversibly as well as adiabatically the process is isentropic as shown in Figs. 26.3 and 26.4. Because the process is adiabatic $q = 0$ and the steady-flow energy equation becomes

$$-w_{xrp} = (h_1 - h_4)$$

or
$$w = w_{xp} = -(h_1 - h_4) \tag{26.5}$$

26.3 A heat pump

A reversed heat engine may be used to take in energy from a reservoir at a low temperature and to deliver energy to a reservoir at a higher temperature. The low-temperature reservoir could be a river, and the high-temperature reservoir could be a block of offices that require heating. From the expression given in equation (6.7) the performance would be

$$\frac{\text{What one gets that is useful}}{\text{What one pays to get it}}$$

In the case of heating a block of offices what one gets that is useful for this purpose is q_1', and what one pays to get it is w'. It is assumed that the energy q_2' coming from, say, the river is available at no cost and so it is not included in what one pays. Performance in the case of a heat pump is assessed by what is called coefficient of performance, c_{hp}, the suffix hp signifying heat pump. Therefore, from equation (6.7),

$$c_{hp} = \frac{q_1'}{w'} \tag{26.6}$$

If the heat pump were reversible, all the components in the cycle would have to be reversible. For this to be so for the two heat actions, then in Fig. 26.3,

$$T_3 = T_L \quad \text{and} \quad T_1 = T_H$$

For the reversible case both the compression and expansion processes would be isentropic. A throttle could not then be used because, as explained in section 26.2, a throttle can never be reversible. Therefore the work energy w' required to drive the reversed engine would be equal to the work energy w_r that would be given out by the forward working engine if the engine were reversible and if $q_1' = q$. Such a reversed engine would be working on a Carnot cycle as shown by $1-2-3'-4-1$ in Fig. 26.3. The efficiency η_r of such a reversible engine working forwards would be given by,

$$\eta_r = \frac{w_r}{q_1} \tag{26.7}$$

and its coefficient of performance as a heat pump would be

$$c_{rhp} = \frac{q_1}{w_r}$$

$$= \frac{1}{\eta_r} \tag{26.8}$$

the value η_r being the Carnot efficiency of equation (12.2). It should be noted that only in a reversible engine reversed does equation (26.8) apply. In an irreversible engine

$$c_{hp} \neq \frac{1}{\eta}$$

26.4 A refrigerator

It has already been stated that a reversed heat engine may be used to take in energy by heat from a reservoir at low temperature and deliver energy by heat to a reservoir at a higher temperature. The low-temperature reservoir could be foodstuffs or other perishable items that must be stored below room temperature, and the high-temperature reservoir could be a cooling-water supply or the atmosphere itself. In either case, in terms of Fig. 26.1, what one gets that is useful is called the **Refrigeration effect** q_2'. This is the heat energy q_2' transferred from the low temperature reservoir and what one pays to get it is w'. It is assumed that the energy q_1' being transferred to the cooling water or to the atmosphere is wasted. Performance in the case of a refrigerator is called coefficient of performance, c_{ref}, the suffix ref signifying refrigerator. Therefore, from equation (6.7),

$$c_{ref} = \frac{q_2'}{w'} = \frac{q_1' - w'}{w'} \tag{26.9}$$

or $$c_{ref} = c_{hp} - 1 \tag{26.10}$$

from equation (26.6).

If the refrigerator were reversible, then the same criteria would apply as did for the heat pump in section 26.3:

$$\text{In Fig. 26.3 } T_3 = T_L \quad \text{and} \quad T_1 = T_H$$

and the turbine and compressor processes must be isentropic. For the reversible engine working backwards

$$c_{r\,ref} = \frac{q_{r2}}{w_r} = \frac{q_1 - w_r}{w_r}$$

$$= \frac{1}{\eta_r} - 1 \tag{26.11}$$

It should again be noted that only in a reversible engine do equations (26.8) and (26.11) apply when they relate the coefficients of performance of the heat pump and refrigerator to the efficiency of the engine. In the case of practical heat pumps and refrigerators equations (26.6), (26.9) and (26.10) should be used.

26.5 Properties of suitable working fluids

In a system with a working fluid undergoing a liquid-vapour refrigeration cycle, what are the desirable properties of the ideal working fluid? These desirable properties are:

(a) The critical temperature should lie above atmospheric temperature so that condensation may occur within the condenser.

(b) The specific enthalpy per degree of the liquid and vapour should be low so that the work required to be done in the compressor is a minimum.

(c) The enthalpy of evaporation (or condensation) should be high so that the flow rate of the working fluid is low and the plant size may be minimized.

(d) It should be arranged so that the lowest vapour pressure of the fluid in the cycle is greater than atmospheric pressure so there is no tendency for air to leak into the system.

(e) The fluid should be such that it is chemically inert to common materials so that the plant may be constructed cheaply.

(f) The fluid should have a low toxicity in order that a leak will cause the smallest hazard.

Although no fluid is perfect, the most suitable fluids are a family of organic chemicals called 'the Freons'. Other fluids used in the past include carbon dioxide, sulphur dioxide, ammonia and methyl chloride, the properties of which can be obtained from the tables of Ref. (3).

Reversed heat engines (Q and A)

Q.1. A refrigerator uses ammonia as its working fluid or refrigerant. What is its coefficient of performance if it were reversible and if its working pressures were $11·67 \times 10^5$ N/m^2 and $3·691 \times 10^5$ N/m^2?

A.1. From Ammonia property tables it can be seen that the temperatures in the condenser and in the evaporator would be 30°C and –4°C respectively. The efficiency of a reversible engine working between these two temperatures would be given by equation (25.6) as,

$$\eta_r = \frac{30 + 273 - (-4 + 273)}{30 + 273}$$
$$= 0·112$$

From equation (26.11)

$$c_{r\,ref} = \frac{1}{0·112} - 1$$
$$= 7·93$$

Q.2. In the reversible refrigerator of Question 1 if the rate of supply of energy to the pump were 2 kW what would be the refrigeration effect over 10 min?

A.2. W' is given as 2 kW and the coefficient of performance from Question 1 equals 7·93 so substituting back in the 'rate' version of equation (26.11)

$$\dot{q}'_{r2} = 7·93 \times 2 \text{ kW}$$

Then in 10 min the refrigeration effect would be

$$\dot{Q}_{r2} = 7·93 \times 2 \times 60 \times 10 \text{ kJ}$$
$$= 9\,516 \text{ kJ}$$

Q.3. If the refrigerator described in Question 1 were not reversible what information would one require to find its coefficient of performance?

A.3. Equation (26.9) would be used to find c_{ref} and so two values from q'_2, w' and q'_1 would be required.

Q. 4 If a heat pump used Freon 12 as its working fluid and worked between temperatures of 10°C and 25°C what would be the dryness of the freon leaving the adiabatic throttle. (See Fig. 26.7)

A. 4. $h_2 = 60$ kJ/kg from tables
            $= h_3$

but $h_2 = 45 + \sigma_3 \, 147$

so $\sigma_3 = 0\cdot102$

26.6 Summary

A reversed heat engine has been described and the function of its components—condenser, throttle, evaporator and compressor—have been studied. The use of the reversed heat engine as both a heat pump and as a refrigerator have been discussed and the coefficients of performance of each of these defined. The properties of an ideal working fluid have been outlined. Because we are dealing with a working fluid operating in the two-phase region, property tables must be used as they are reversed vapour power plants.

26.7 Questions for the reader

Q. 1. Draw a T-s diagram for a liquid-vapour refrigerant cycle working between two reservoirs at temperatures of —15°C and +30°C where

(a) the cycle is reversible,
(b) the cycle has a throttle as the expander but the adiabatic compressor is reversible,
(c) the cycle is a practical cycle, but with reversible heat exchangers.

[See Fig. 26.6]

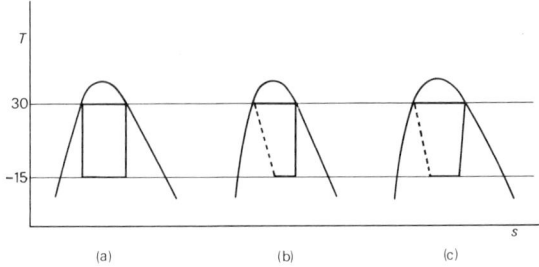

FIG 26.6   Answer to question 1

Q.2. A refrigerator has ammonia as its working fluid working between −15°C and +30°C. Compare the coefficients of performance, if the cycle (a) operates on a reversed Carnot cycle, and (b) operates on a cycle similar to the one in Question 1 (b).

[(a) 5·73, (b) 5·07]

Q.3. Why do the answers to (a) and (b) of Question 2 differ?

[Cycle (a) is reversible but cycle (b) has a throttle in it which is inherently irreversible]

Q.4. For the two cycles of Question 2 what is the refrigeration effect from each case?

[(a) −977, (b) −960 kJ/kg]

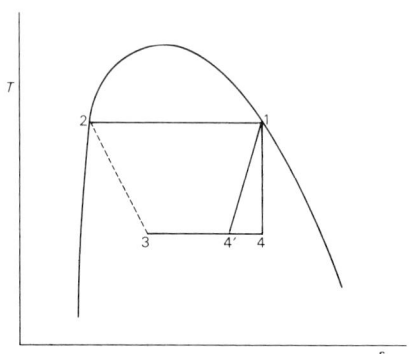

FIG 26.7   Two cycles of question 4

Q.5. In a refrigerator using freon it is arranged that the vapour will enter the compressor as a dry saturated vapour. If the refrigerator operates between a pressure of $1·00 \times 10^5$ N/m^2 and a pressure of $7·50 \times 10^5$ N/m^2, what is the refrigerator's coefficient of performance?

[3·03]

Q.6. A man requires to heat his house. He calculates that he requires a system capable of a power output of 15 kW. He is told of a heat pump using a 5 kW compressor which has 50 per cent of the performance of an equivalent reversible pump. The system is required to operate between −13°C and +27°C. Is this heat pump likely to solve his problem?

> [Yes, calculations show for heat pump and these conditions,
> $w_{xp} = 4$ kW]

Q. 7.   The consulting firm who are advising on the design of a refrigeration scheme for a warehouse, calculate the energy loss from the warehouse, which is maintained at −20°C, as 1 MW. They propose to build a series of heat pumps linking the warehouse and a nearby block of flats. The flats must be maintained at 27°C and each flat has a 10 kW power requirement. How many flats can be linked to the warehouse if reversible heat pumps are used?

> [118]

Q. 8.   An office block is maintained at 22°C by the use of a heat pump which uses a river as the lower temperature reservoir. The plant is driven by a 100 kW compressor on a cycle similar to that of Question 1 (b) using ammonia as the working fluid. If it is known that the energy loss from the building is 250 kW, what is the lowest temperature to which the river may drop?

> [0°C; at this temperature the river freezes! At this temperature and below the heat pump is unsatisfactory]

27

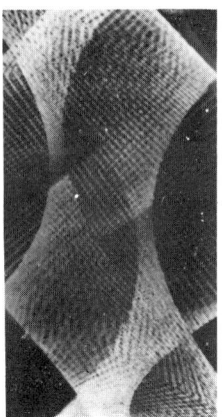

Improvements and the future

So far in this book we have introduced the reader to elementary thermodynamic theory and we have shown how the theory can be used to predict the behaviour of thermodynamic cycles and the processes those cycles comprise. This is by no means the end of the story: but only its beginning because thermodynamics can predict the behaviour of more complicated and frequently used cycles. Lines along which the subject as now known has further developed are outlined in this last chapter.

27.1 Improvements to gas power cycles

In Fig. 22.4 and again in Fig. 27.1 the state path of the working fluid in a simple gas power plant is shown by the cycle 1-2-3-4-1. The state path in the ideal Carnot cycle operating between the same maximum and minimum temperatures is 1″-2-3″-4-1″. In the power plant the state of the gas follows the constant pressure paths 1-2 and 3-4 instead of 1″-2 and 3″-4 of the ideal cycle because the components in which a gas undergoes a simple constant-pressure heat-transfer process are a good deal less complex than those in which it would undergo an isothermal simple heat-transfer process.

It can be seen in Fig. 27.1 that the gas at 3 when it leaves the turbine is at a higher temperature than the gas at 1 when it leaves the compressor. The hot gases at state 3 can be used to heat the relatively cool gas at 1 from T_1 to a higher temperature. The theoretical limit to which gas leaving the compressor could be heated by gas leaving the turbine is T_x of Fig. 27.1 where $T_x = T_3$. Of course, while heating the gas leaving the compressor the gas leaving the turbine undergoes a reduction in temperature. The theoretical limit to which gas leaving the turbine can be cooled by the gas leaving the compressor is T_y of Fig. 27.1 where $T_y = T_1$. Therefore by arranging a counter flow system within a heat exchanger, like the heat exchanger in Fig. 27.3, it is theoretically possible for the temperature of all the gas leaving the compressor to be raised to T_x while the

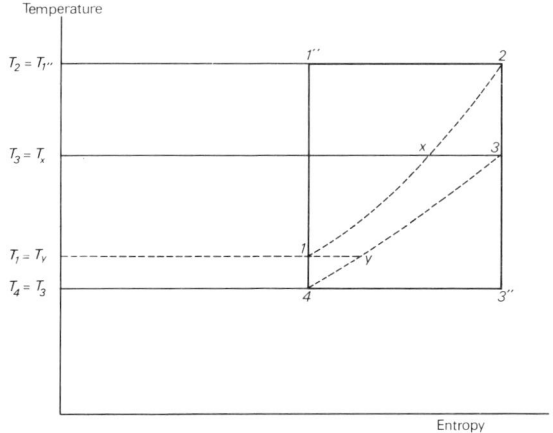

FIG 27.1   Gas-power cycle without and with heat exchanger

temperature of all the gas leaving the turbine falls to T_y, assuming that their specific enthalpies per degree are constant. The gas is in states $1, x, 2, 3, y$ and 4 respectively of Fig. 27.1 when at points $1, x, 2, 3, y$ and 4 of the flow diagram shown in Fig. 27.3.

      The work ratio (equation (22.2) is unchanged by the introduction of a heat exchanger of the type mentioned in the last paragraph. The work ratio in a gas-turbine power plant is improved by an increase of turbine and a reduction of compressor work. The distance measured along an isentropic line between two constant-pressure lines, for example a comparison of the distances 1-4 and 2-3 of Fig. 27.1 shows a decrease for lower values of entropy. This leads one to the

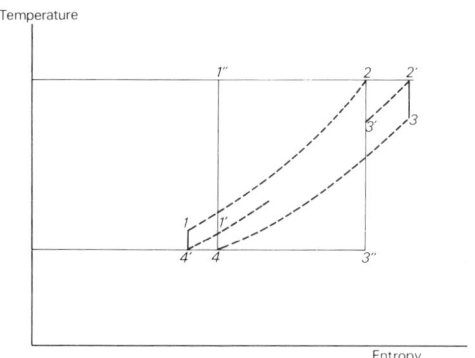

FIG 27.2   Gas-power cycle with intercooling and reheating

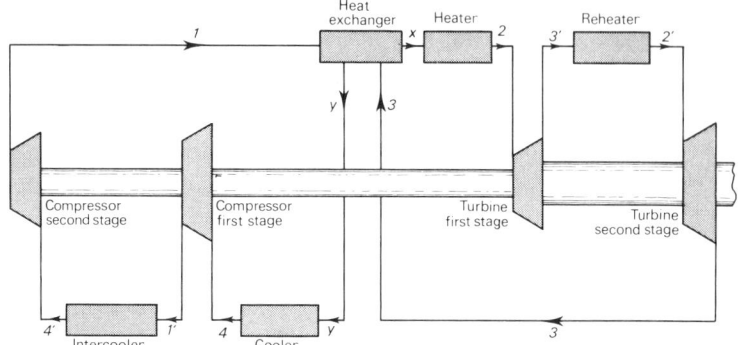

FIG 27.3   Some improvements incorporated in a gas power cycle

conclusion that the further one moves conditions to the left in Fig. 27.1 the less compressor work one has to have done. For the same overall pressure range a reduction in compressor work has been achieved in the way shown in Fig. 27.2 by only partly compressing the gas from 4 to 1′ in a first-stage compressor and cooling the gas in an intercooler before admitting it to the second-stage compressor in which the state of the gas is brought to its final compressed state at 1. The gas is in states 4, 1′, 4′ and 1 respectively of Fig. 27.2 when at points 4, 1′, 4′ and 1 of the flow diagram of Fig. 27.3. Similarly, the further to the right in Fig. 27.2 one moves conditions around the turbine the more turbine work will be done. For the same overall pressure range an increase in turbine work has been achieved in the way shown in Fig. 27.2 by only partly expanding the gas from 2 to 3′ in a first stage turbine and reheating the gas in a reheater before admitting it to the second-stage turbine in which the gas is expanded to its final expanded state at 3. The gas is in states 2, 3′, 2′ and 3 respectively of Fig. 27.2 when at points 2, 3′, 2′ and 3 of the flow diagram of Fig. 27.3.

27.2 Improvements to vapour power cycles

In Fig. 25.1 and again in Fig. 27.4 the state path of a vapour in a power plant working on an ideal Rankine cycle is shown by the cycle $1'-1-2-2_S-3-4-1'$ and the ideal Carnot cycle between the same maximum and minimum temperatures is shown in Fig. 27.4 by the state path $1''-2_S-3-4-1''$.

It can be seen in Fig. 27.4 that the energy q_1 is transmitted to the working fluid at a temperature that varies from $T_{1'}$ to T_{2S}. This is the reason why the plant, working on a simple ideal Rankine cycle, has an efficiency substantially below the Carnot efficiency. If the expansion in the turbine is in two or more stages with a reheater between the stages as shown in Fig. 27.5 three improvements result,

(a) The dryness of the vapour in the turbine is substantially reduced —compare the state point 3″ with 3.

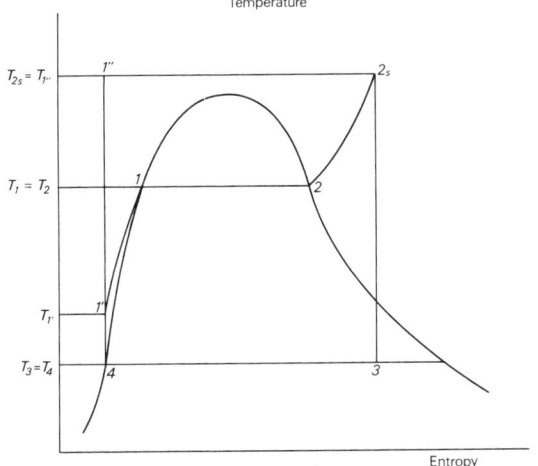

FIG 27.4   Rankine and Carnot cycles

(b) Heat energy q_1 is transferred more isothermally—compare the mean temperature of $1'$-2_s together with $2'$-$2'_s$ and $2''$-$2''_s$. This increases the efficiency.

(c) Work energy is increased—compare the areas giving q_1 and q_2. The work energy is the difference between these.

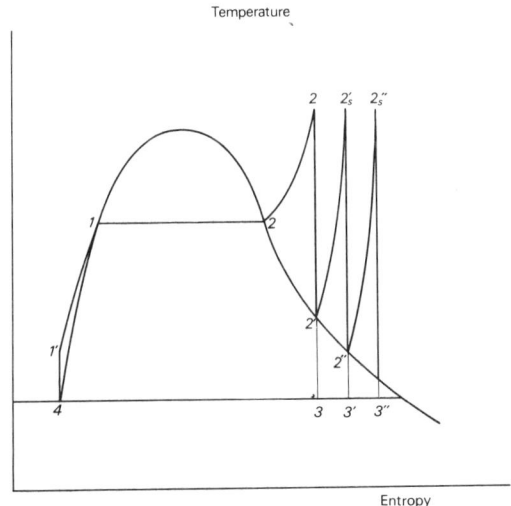

FIG 27.5   Rankine cycle with reheating.

FIG 27. 6   Part of one of the reactors at Dungeness 'A' nuclear
power station. *(Photography by courtesy of the Central
Electricity Generating Board.)*

It has already been stated that one reason for inefficiency in the cycle of Fig. 27. 4 which is also present in the cycle of Fig. 27. 5 is the fact that q_1 is not taken in isothermally. This is particularly evident in the heating from 1' to 1 where the fluid is liquid and its temperature is rising rapidly. If the energy required to raise the fluid's temperature from state 1' to state 1 could be taken reversibly from the fluid somewhere else in the cycle the raising of the liquid's temperature from 1' to 1 would not be part of the energy q_1. The energy q_1 would then be delivered to the engine along the lines $1\text{-}2\text{-}2_S$, $2'\text{-}2'_S$, and $2''\text{-}2''_S$, which is more nearly isothermal. Taking energy from the fluid somewhere else in the cycle can be done, but of course not quite reversibly. For instance it can be taken from the fluid in the turbine in one of several ways. These processes are regenerative heating.

FIG 27. 7   A Parsons 660MW turbine line under construction.
(Photograph by courtesy of C.A. Parsons and Co. Ltd.)

From what has been written above it should be evident to the reader
that modern power plants are highly complex. Part of the top of a nuclear reactor
and the crane for removing spent nuclear fuel used in a modern power station is
shown in Fig. 27.6. The whole reactor itself is only a part of what we have hitherto called a boiler and furnace. In Fig. 27.7 a modern steam turbine is shown during
assembly. In the figure one of three low-pressure shafts is being lowered into
position (such a low-pressure shaft would be within the second stage of the turbine
shown in Fig. 27.3). The low-pressure shaft of Fig. 27.7 has a mass of 47 000 kg.
A large modern surface condenser under construction is shown in Fig. 27.8. Two
of these condensers operating in conjunction with a 500 MW turbine-generator set
are arranged, one on each side of the low-pressure turbine casing. Each condenser
will handle 7 500 kg/s of cooling water and the 14 000 m^2 of cooling surface is provided by over 9 000 tubes, each of about 25 mm diameter.

FIG 27.8   A large surface condenser under construction. *(Photograph by courtesy of C.A.Parsons and Co.Ltd.)*

27.3 The future of combustion

When early man first stumbled upon fire and learnt that his wood-land—that is to say the woodland that he saw around him—could be made to give up its energy by burning, he would have been gratified to know, if such information could have been made available to him, that the natural control of population by war and disease and the rate of growth of new woodland would ensure sufficient wood as fuel for his lifetime and even beyond.

Modern man when he stumbled on the possibility of using nuclear fuel became aware of many facts about his existence in the world. For instance, war had become so dangerous that he was afraid for his own safety. And the exhaustion of the traditional fuels became a frightening possibility. The consumption of oxygen, our precious life giver, increases at a frightening rate with increases of human population, and an overall increase in living standards. All these fears gathered inside him and collectively gave him the complex situation in which he finds himself today. In the same way anyone with the patience to have read through this book will in making decisions have to relate other subjects to this topic in solving problems.

With regard to a deterioration in the quality of the atmosphere there are sound reasons for man's fears. The oxygen is being combined with carbon and to a lesser extent with other elements at an increasing rate. This irreversible burning will have to be discontinued and other forms of energy release found, from which unpleasant consequential effects can be removed. One unpleasant effect is the residual activity in nuclear fuel. Means must be found of arresting this activity in nuclear fuel. Means must be found of arresting this activity before the activated substance is discarded.

When the first steam railways had been built men who lived near the new railway lines were quite sure that the rainfall had increased in the areas on both sides of the lines and that the steam from the locomotives would turn the areas into swamps. It has not yet happened nor do we any longer think it likely to happen. Perhaps other fears like our fears of pollution and of waste will prove equally groundless and perhaps the second law of thermodynamics will eventually decide all matters when the level of availability of energy on the planet is too low to sustain life.

Index